Planning and Control of Maintenance Systems

Salih O. Duffuaa · A. Raouf

Planning and Control of Maintenance Systems

Modelling and Analysis

Second Edition

Springer

Salih O. Duffuaa
Department of Systems Engineering
King Fahd University of Petroleum
and Minerals
Dhahran
Saudi Arabia

A. Raouf
University of Management and Technology
Lahore
Pakistan

ISBN 978-3-319-19802-6 ISBN 978-3-319-19803-3 (eBook)
DOI 10.1007/978-3-319-19803-3

Library of Congress Control Number: 2015941354

Springer Cham Heidelberg New York Dordrecht London

Printed on acid-free paper

Springer International Publishing AG Switzerland is part of Springer Science+Business Media
(www.springer.com)

To my wife Samia, daughter Sara, and sons Mohammed and Omar

—Salih O. Duffuaa

To my wife Razia

—A. Raouf

Preface

Maintenance systems play a key role in achieving organizations' missions and abilities to attain their profit targets and survive in globally competitive marketplace and changing economies. The purpose of this book is to present maintenance as an integrated system with objectives, strategies, and processes that need to be planned, designed, engineered, and controlled using statistical and optimization techniques. The theme of the book is the strategic holistic system approach for maintenance. This approach enables maintenance decision makers to view maintenance as a provider of a competitive edge not a necessary evil. The book has fourteen chapters including maintenance systems, maintenance strategic and capacity planning, planned and preventive maintenance, work measurements and standards, material (spares) control, maintenance operations and control, planning and scheduling, maintenance quality, training, reliability-centered maintenance (RCM), total productive maintenance (TPM), intelligent maintenance systems, maintenance performance, productivity, and continuous improvement.

Chapter 1 provides an over view of the book. It outlines the philosophy and approach the book adopts in addressing maintenance systems. This chapter presents maintenance as a system with inputs, outputs, processes, and activities that need to be planned, designed, engineered, executed, and controlled. In addition, it highlights maintenance activities and strategies and concludes with maintenance terms and their definitions.

Chapter 2 focuses on maintenance strategic and capacity planning. Forecasting techniques that are useful for maintenance planning are presented. The need to align maintenance strategic plan with organizational goals is emphasized. Heuristic and mathematical models for capacity planning are discussed and demonstrated by examples from maintenance.

Chapter 3 addresses preventive maintenance and breakdown repairs. Various useful diagnostic techniques for monitoring equipment health and condition are presented. These techniques play a key role in designing condition-based maintenance. The steps for designing an effective planned maintenance are discussed. Mathematical models for determining optimal preventive maintenance and inspection frequencies are presented. The concept of imperfect maintenance is explained, and approaches for modeling its effect are provided. Finally, a brief description of delay time modeling for determining preventive maintenance is provided.

Chapter 4 discusses the need for maintenance job time standards and covers techniques for developing such standards. The techniques covered include work measurement, work sampling, standard data, and comparative estimation.

Chapter 5 focuses on spare parts provisioning for maintenance. Effective ordering polices for repairs and ordering strategies are discussed. A methodology for estimating material and spare parts costs is outlined.

Chapter 6 discusses maintenance operation and control systems. The role of the work order system is presented in detail. Guidelines for designing effective work order systems are provided. The components of the maintenance control system and their underlying structures are outlined. Work order coordination, maintenance unit control and reports for feedback and improvement are presented as well.

Chapter 7 presents maintenance planning and scheduling. The elements of medium- and short-term planning are discussed. Effective planning models for scheduling including networks and critical path analysis are presented with examples. This chapter concludes with planning turnaround maintenance.

Chapter 8 describes maintenance quality control. The responsibility, organization, and tools for controlling and improving quality of maintenance work are presented with examples.

Chapter 9 deals with maintenance training and discusses skills required for maintenance. In addition, this chapter presents a framework for designing a training program for maintenance personnel and outlines means for evaluating and improving the effectiveness of such programs.

Chapter 10 addresses computerized maintenance systems and presents the requirements for a typical maintenance management information system (MMIS). It also provides a methodology for evaluating MMISs.

Chapter 11 discusses reliability centred maintenance (RCM) and presents a step-by-step approach for developing a RCM program. In addition, this chapter outlines the steps for implementing RCM.

Chapter 12 presents total productive maintenance (TPM) as a successful approach for managing maintenance. This chapter provides the main elements of TPM and a road map for implementing it.

Chapter 13 addresses the concepts and techniques of e-maintenance. The development of information and communication technology made it possible to go beyond predictive maintenance to e-maintenance. This chapter presents the elements of e-maintenance and the requirement for its implementation together with few case studies.

Chapter 14 focuses on maintenance performance measures, productivity, and continuous improvement. It presents maintenance performance measures, indices, quality improvement tools, root cause analysis, benchmarking, and process re-engineering. The applicability of these techniques in maintenance is demonstrated using case studies.

Each chapter includes a number of exercises, which are of two types. Answers to the first type of exercises can be found within the text. Answers to the second type require some research. The authors have tried as much as possible to make this book self-contained. Techniques and models used in the text are explained within the text; however, for readers who did not have a basic course in probability and statistics, an appendix is provided to alleviate this deficiency. The book has a modest mathematical level. Engineering or management students, practicing engineers, and managers who have completed an introductory course in statistics should have no difficulty understanding almost all of the text. It is our intent to give readers an understanding of the relevant methodology and how to apply it, rather than to provide a complete treatment of the mathematical theory.

This is the second edition of the book. Four new chapters are added to the second edition and three chapters are revised substantially to reflect developments in maintenance since the publication of the first edition. The new chapters cover reliability-centered maintenance, total productive maintenance, e-maintenance and maintenance performance, productivity, and continuous improvement. The revised chapters include Chap. 1 the introduction, Chap. 2 *Maintenance Strategic and Capacity Planning*, and Chap. 14 *Maintenance Performance, Productivity, and Continuous Improvement*. The write-up of the other chapters has been improved to enhance presentation and readership.

This book is suitable as a textbook or a reference for professionals and practitioners. It can be used as a text for undergraduate- or first-year graduate-level courses in maintenance. It will be of interest to industrial engineering, mechanical engineering, electrical engineering, and industrial management students. It can also be used as a textbook for short courses on maintenance in industry.

Acknowledgments

We would like to express our grateful appreciation to the organizations and the individuals who contributed to this book. The excellent facilities provided by King Fahd University of Petroleum and Minerals and University of Management and Technology (UMT) Lahore, Pakistan, for completing this project are acknowledged.

The first author acknowledges the support provided by King Fahd University of Petroleum and Minerals, Dhahran, Saudi Arabia, under the Project No. IN141007 without this support the completion of this book on time would not have been possible.

We appreciate the comments and valuable suggestions received from many colleagues including Prof. M. Ben Daya, Prof. S. Selim, and Dr. A. Haroun. The editorial assistance of Dr. B. Bremberg is highly appreciated, and the efforts of Mr. Syed Abdulwaheed Shah for typing the manuscript are acknowledged.

Last but not least, we would like to express our gratitude to our patient supportive families who suffered the loss of many weekends during the preparation of the second edition of the book, including Duffuaa's family wife Samia, daughter Sara, and sons Mohammed and Omar, and Raouf's wife Razia.

Contents

About the Authors

Dr. Salih O. Duffuaa is a professor of Operations Research and Industrial Engineering at the Department of Systems Engineering, King Fahd University of Petroleum and Minerals, Dhahran, Saudi Arabia. He received his B.Sc. (Honors) in Mathematics and a Diploma in Statistics from the University of Khartoum, Sudan. He obtained his Ph.D. in Operations Research from the University of Texas at Austin in 1982. His current research interests are in mathematical modeling, network programming, maintenance systems optimization, and quality control. He has published extensively in these areas. His publications appeared in the Journal of Optimization Theory and Applications, European Journal of Operational Research, Journal of the Operational Research Society, International Journal of Production Research, Applied Mathematical Modeling, Journal of Quality in Maintenance Engineering, and others. He is on the editorial board of the International Journal of Industrial Engineering, and he is the editor of the Journal of Quality in Maintenance Engineering. He is a certified supply chain professional and has consulted to industry on many projects related to maintenance, quality, and optimization.

Dr. A. Raouf is a distinguished national professor and professor emeritus University of Management and Technology (UMT) Lahore, Pakistan. He has retired from active teaching but still involved in consulting and supervising research students. He was a university professor at UMT for more than a decade and prior to that he was a rector in Ghulam Ishaq Khan Institute (GIKI), Pakistan. He spent more than 10 years at KFUPM, Saudi Arabia. He spent the years 1966–1984 at the University of Windsor, Canada; out of this period, he was the chairman of the Industrial Engineering Department for 7 years. His contributions to the profession include more than hundred articles in reputed journals and 12 books (authored or edited). He is on the editorial board of quite a few professional journals and the editor-in-chief of JQME.

Chapter 1
Maintenance Systems

1.1 Introduction

In an era of competitive marketplace and ever-changing economics, there are intense pressures on organizations in both the manufacturing and service sectors to respond to customers' demands and deliver high-quality products in a timely manner. These lead organizations to gear their strategies toward agility, quality, automation, and high performance. This has resulted in very high investments in equipment and people. To achieve the targeted rates of return on investment and survive in these dynamic economics, equipment has to be reliable and safe to operate without costly work stoppages and repairs. Many manufacturing companies have adopted just-in-time (JIT) lean programs and are operating with work-in-process so low that there is no inventory reserve to use in case equipment unavailability occurs. The above developments brought forward the role of maintenance as a key activity in manufacturing and service organizations.

The theme of this book is the holistic system approach for maintenance and its link to organizations' strategies and objectives. This approach enables maintenance decision makers to project maintenance as a provider of a competitive edge not a necessary evil. The consequence of this theme is that maintenance is viewed as an integrated system with objectives, strategies, processes, and activities that need to be aligned with organization's objectives, planned, designed, engineered, and controlled using statistical and optimization techniques. This chapter provides an overview and highlights the key features of the book. In addition, it presents the components of a maintenance system that needs to be planned, organized, and optimized, in order to maximize the output of a maintenance system and achieve organization objectives with the best utilization of resources. The rest of the chapter is organized as follows: Sect. 1.2 introduces the strategic system approach to maintenance followed by those elements that need to be planned in Sect. 1.3. Section 1.4 focuses on organizing and designing activities, and Sect. 1.5 presents controlling activities. Section 1.6 outlines the concept of managing for quality

S.O. Duffuaa and A. Raouf, *Planning and Control of Maintenance Systems*,
DOI 10.1007/978-3-319-19803-3_1

followed by reliability-centered maintenance as a strategy for optimizing preventive maintenance in Sect. 1.7. Section 1.8 presents total productive maintenance as an approach for managing maintenance, and Sect. 1.9 addresses human behavior in maintenance. Section 1.10 outlines the elements of intelligent maintenance systems followed by definitions of common terms in maintenance management in Sect. 1.11. Section 1.12 summarizes the chapter.

1.2 Strategic Holistic System Approach

Maintenance is defined as the combination of activities by which equipment or a system is kept or restored to a state in which it can perform its designated function. It is an important factor in product and service quality and can be used as a strategy for successful competition. Inconsistencies in production equipment operation result in excessive variability in the product and thus cause defective output. To be able to produce a high level of quality, production or service equipment must operate within specifications that are attainable by timely maintenance actions.

A system is a collection of components that work together toward a common objective(s). Maintenance can be considered as a system with a set of processes and activities carried out in parallel with production or service systems. A diagrammatic relationship among organizational objectives, production/operation process, and maintenance is shown in Fig. 1.1. Production/operation systems are usually concerned with converting inputs such as raw materials, manpower, and processes into products/services that satisfy customer needs. The primary outputs of the production/operation systems are finished products or services, and the secondary output is degraded or failed equipment. This secondary output generates demand for maintenance. The maintenance system takes this as an input and adds to it know-how, manpower, and spares, and produces equipment/facilities in good operating condition, that provide capacity for production or service.

The overall primary goal of a production or a service system is to maximize profit from the available market opportunities, and its secondary goal is concerned with economical and technical aspects of the conversion process. Maintenance systems assist in achieving these goals as well, by increasing profits and customer satisfaction. These are achieved by minimizing the plant downtime, improving the quality, increasing the productivity and by reliable timely delivery of orders to customers. Production and service systems have been optimized as an integrated system and have been studied quite extensively compared to maintenance systems. Possible reasons include the following: (1) Traditionally, maintenance has been regarded as a necessary evil and at best a system driven by production, (2) maintenance in an organization has complex relationships with other functions, and (3) the output of maintenance is hard to measure and quantify.

The role of maintenance systems has been long realized in manufacturing organizations; however, it has become clear that the functions of maintenance are

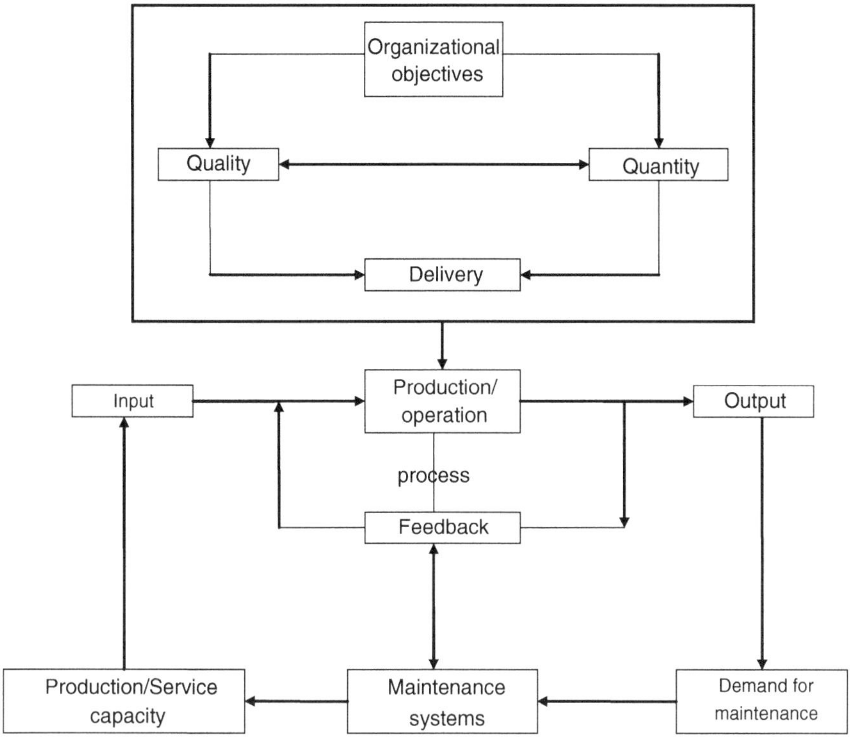

Fig. 1.1 Relationships between maintenance systems and organization objectives

also essential for service organizations such as hospitals, banks, educational institutions, and department stores. In organizations such as hospitals, for example, X-ray and brain scanning machines must be kept up all the time due to their criticality to human life. The concepts, models, and techniques in this book that are used for planning, designing, organizing, and controlling of maintenance systems are applicable to all organizations that perform a business function. Therefore, the reader should be aware that there is a broad spectrum for the use of the material in this book.

A maintenance system can be viewed as an integrated input–output model. The inputs to such a model are labor, management, tools, spares, equipment, etc., and the output is equipment that is up, reliable, and well configured to achieve the planned operation of the plant. This enables us to optimize the resources for maximizing the output of a maintenance system. A typical maintenance system is shown in Fig. 1.2. Activities needed to make this system efficient and effective include planning, designing, organizing, controlling, and improving that are shown in the figure.

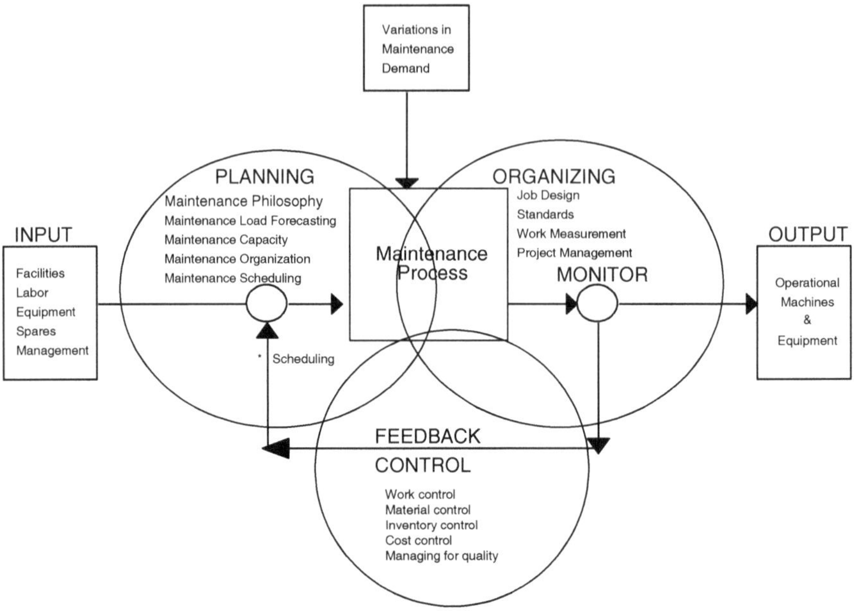

Fig. 1.2 Maintenance systems

1.3 Planning Activities

Planning activities generally include the following Martin [8]:

1. Strategic systems alliances
2. Maintenance strategies
3. Maintenance load forecasting
4. Maintenance capacity
5. Maintenance organization
6. Maintenance scheduling.

A description of each of these activities is given in the next subsections.

1.3.1 Strategic Systems Alliance

Maintenance departments should have their own strategic plans that are aligned with their organizations' strategic objectives and managed with maximum efficiency and effectiveness taking into consideration the holistic system approach. Strategies for maintenance operations should be selected among alternatives to achieve organizations' objectives. Recent works of Tsang [11, 12], Murthy et al. [9], and

Al-Turki [1] have discussed issues related to maintenance strategic planning and identified several important issues that are essential in deciding maintenance strategic plan. These issues include maintenance outsourcing, organization, methodology, and support. At the strategic level, management decides the organization's strategic goals, and from these goals, the maintenance unit goals are derived to support the organization's strategic goals and mission. Then, the right organization, maintenance methodology, and support are selected to achieve these goals. Maintenance methodology is the strategy to be used at the equipment maintenance level, and the support includes maintenance key processes, manning, information system, training, and performance management and reward system. Maintenance strategic planning is covered in Chap. 2.

1.3.2 Maintenance Strategies

The maintenance methodology is derived from the maintenance system objectives that are aligned with the organization's mission and strategic goals. To implement the methodology, the following strategies can play an effective role if applied in the right mix and fashion. The strategies are as follows:

1. Breakdown/corrective maintenance
2. Preventive maintenance
   - 2.1 Time- or use-based preventive maintenance
   - 2.2 Condition-based preventive maintenance
3. Opportunity maintenance
4. Fault finding
5. Design modification
6. Overhaul
7. Replacement
8. Reliability-centered maintenance
9. Total productive maintenance

1.3.2.1 Breakdown/Corrective Maintenance

This type of maintenance is only performed when the equipment is incapable of further operation. There is no element of planning for this type of maintenance. This is the case when the extra cost of other types of maintenance cannot be justified. This type of strategy is sometimes referred to as run to failure strategy. It is applicable mostly to electronic components.

1.3.2.2 Time- or Use-Based Preventive Maintenance

Preventive maintenance (PM) is any planned maintenance performed to counteract potential failures. It could be implemented based on the use or equipment condition. The time- or use-based PM is performed on an hours run or calendar basis. It needs a high level of planning. The specific routines to be carried out are known as well as their frequencies. In determining the frequency, usually knowledge about failure distribution or equipment reliability is needed.

1.3.2.3 Condition-Based Preventive Maintenance

Condition-based maintenance is carried out on the basis of the known condition of the equipment. The condition of the equipment is determined by monitoring key equipment parameters whose values are affected by the condition of the equipment. This strategy is also known as predictive maintenance. Techniques for preventive maintenance are covered in detail in Chap. 3.

1.3.2.4 Opportunity Maintenance

This type of maintenance, as the name implies, is carried out when the opportunity arises. Such opportunities may arise during shutdown periods of a particular system and can be utilized for carrying out known maintenance tasks.

1.3.2.5 Fault Finding

Fault finding is an act or inspection performed to assess the level of failure set on. An example of fault finding is checking the spare tire of a car prior to taking a long trip.

1.3.2.6 Design Modification

Design modification is carried out to bring equipment to a current acceptable condition. It involves improvement and sometimes manufacturing and capacity expansion. It usually requires coordination with engineering and other departments in the organization to do this type of work.

1.3.2.7 Overhaul

Overhaul is a comprehensive examination and restoration of equipment or a major of equipment to an acceptable condition. This is usually a major task.

1.3.2.8 Replacement

This strategy is to replace the equipment instead of performing maintenance. It could be a planned replacement or replacement upon failure.

Each of the maintenance strategies described above has a role to play in the plant operation. It is the optimal mix of these maintenance strategies that results in the most effective maintenance philosophy. The size of the plant and its planned level of operation coupled with the applicable maintenance strategy can assist in estimating the maintenance load or the desired output of the maintenance system. Figure 1.3 summarizes the maintenance strategies and for more on maintenance strategies see [3, 6].

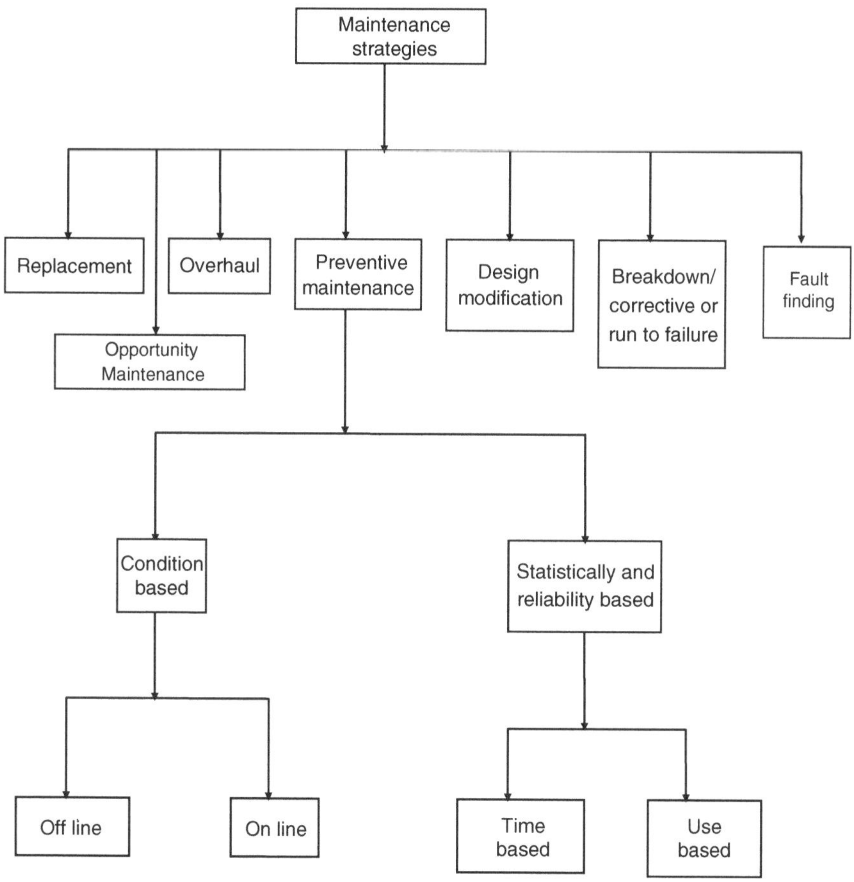

Fig. 1.3 Maintenance strategies

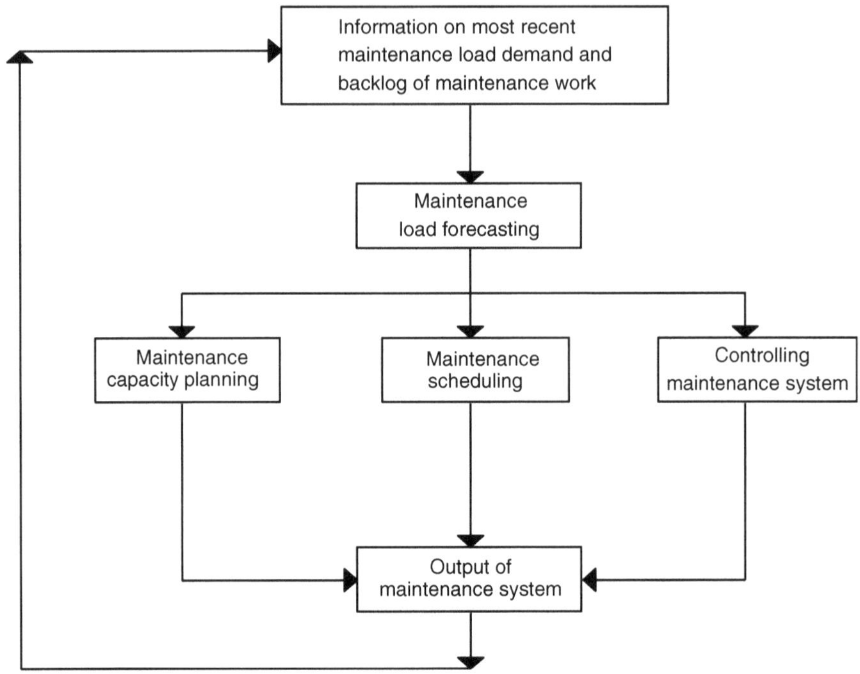

Fig. 1.4 Maintenance load forecasting and maintenance system

1.3.3 Maintenance Load Forecasting

Maintenance load forecasting is the process by which the maintenance load is predicted. The maintenance load in a given plant varies randomly and, among other factors, can be a function of the age of the equipment, the rate of its use, maintenance quality, climatic factors, and skills of maintenance craftsmen. Maintenance load forecasting is essential for achieving a desired level of effectiveness and resource utilization, and without it, many maintenance functions cannot be performed well. The role played by maintenance load forecasting in a maintenance system is shown in Fig. 1.4.

1.3.4 Maintenance Capacity Planning

Maintenance capacity planning determines the resources needed to meet the demand for maintenance work. Those resources include manpower, material, spare parts, equipment, and tools. Critical aspects of maintenance capacity are the numbers and skills of the craftsmen, required maintenance tools, etc. Since the maintenance load is a random variable, the exact number of various types of craftsmen cannot be determined. Therefore, without reasonably accurate forecasts for the future maintenance work demand, it would not be possible to do proper long-range capacity planning. In order to have better utilization of manpower, organizations tend to reduce the number of available craftsmen than their expected need. This is likely to result in the backlog of uncompleted maintenance work. This can be completed by letting the existing craftsmen work overtime or by seeking outside contractor assistance. This backlog can also be cleared when the maintenance load is less than the capacity. This is actually the main reason for keeping a backlog. Making long-run estimations is one of the areas in maintenance capacity planning that is both critical and not well developed. Techniques for maintenance forecasting and capacity planning are presented in Chap. 2.

1.3.5 Maintenance Organization

Many factors influence maintenance organization. The factors include plant size, maintenance load, type of organization, and craftsmen skills. Based on these factors and other, maintenance may be organized on departmental, area, or central basis. Each type of organization has its pros and cons. In large organizations, decentralization of maintenance can produce quicker response time and can allow the craftsmen to become more experienced on the problems of a particular section of the plant. However, the creation of a number of small units tends to reduce the flexibility of the maintenance system as a whole. The range of skills available becomes reduced, and manpower utilization is usually less than that in a centralized maintenance unit. In some cases, a compromise solution is possible which is called a *cascade* system. This system enables production area maintenance units to be linked to the central maintenance unit. Such a system is shown in Fig. 1.5.

1.3.6 Maintenance Scheduling

Maintenance scheduling is the process of assigning resources and manpower to jobs to be accomplished at certain times. It is necessary to ensure that the needed craftsmen, the parts, and the materials required are available before a maintenance task can be scheduled. Critical equipment in a plant, are those whose failure will shut down the production process or endanger human life and safety. Maintenance work pertaining to such equipment is treated on a priority basis and is attended to before any other job is undertaken. Occurrences of such jobs cannot be predicted with certainty, and as such, the schedules for planned maintenance in these instances have to be revised. The effectiveness of a maintenance system is greatly

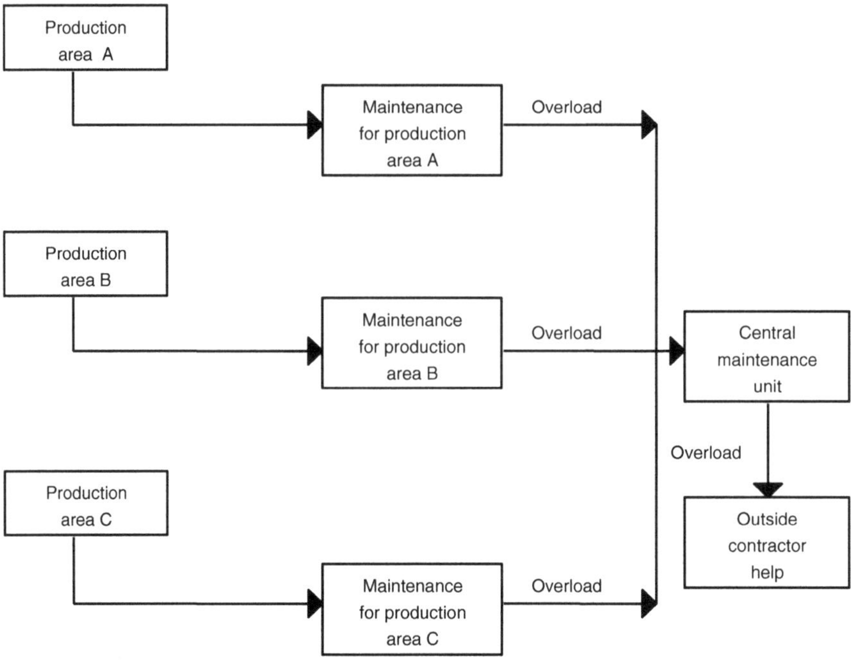

Fig. 1.5 Cascade system

influenced by the maintenance schedule developed and its ability to accommodate changes. A high level of maintenance schedule effectiveness is indicative of a high level of maintenance effectiveness. Chapter 7 covers the necessary tools for effective planning and scheduling.

1.4 Organizing and Designing Activities

Organizing and designing a maintenance system includes the following:

1. Job design
2. Standards
3. Project management.

Maintenance systems are known to be driven by the work load that is issued by the production or operation department as work request. The work requests are planned, executed, and controlled by a work order system. The work orders describe the work, its location, the crafts needed, and the priority of the job. The work control system is presented in Chap. 3.

1.4.1 Job Design

Job design, as related to maintenance work, comprises the work content of each job and determines the method that is to be used, the special tools needed, and the skilled persons required. If the design is standardized, it becomes a standard job.

1.4.2 Time Standards

Once the maintenance task goes through the job design stage, it is necessary to estimate the time needed for completing the job Niebel [10]. Realistic time standards go a long way in monitoring and increasing effectiveness of craftsmen, thus minimizing the plant downtime. It is not essential to have standards for all the maintenance jobs. It is noticed that twenty percent of the maintenance jobs take approximately eighty percent of the time available for maintenance workforce. Efforts should be made to have time standards developed for such time-consuming jobs. It should be obvious that job time standards are needed for maintenance load forecasting, capacity planning, and developing maintenance schedules. Techniques for developing job time standards such as work measurement, work sampling, and comparative estimation are presented in Chap. 4.

1.4.3 Project Management

Planned major overhauls and preventive maintenance service are periodically carried out in most of the large plants. During this period, the entire plant or part of the plant is shut down. It is advantageous to plan and chart the work in order to minimize the downtime and make the best use of resources. Project management involves developing networks of activities and then using techniques such as the critical path method (CPM) or program evaluation and review technique (PERT). Once the network has been developed which includes work breakdown, job sequence, and time estimates for each activity, computer software such Primavera may be used for scheduling the activities and determining the best utilization of resources for more on maintenance planning and scheduling see [4, 5]. A control phase of such a project includes measuring progress regularly, comparing it with the schedule, and analyzing the variance as a percentage of the total work. Corrective actions can be taken to eliminate the shortcomings. Techniques of project management are explained in Chap. 7.

1.5 Control Activities

Control is an essential part of scientific management. Control as applied to a maintenance system includes the following activities:

1. Work control
2. Inventory control
3. Cost control
4. Quality control.

1.5.1 Work Control

The maintenance system is driven by demand for maintenance work. The maintenance work load is greatly influenced by the maintenance strategies. The management and control of the maintenance work is essential for achieving set plans. The work order system is the tool used for controlling the maintenance work. A well-designed work order with a sound reporting system is the heart of the maintenance system. The essential tools for effective control of the maintenance work including the design of a work order are covered in Chap. 6.

1.5.2 Inventory Control

It has been previously stated that for scheduling a maintenance work, it is essential to assure that required spares and material are available. It is physically impossible and economically impractical for each spare to arrive exactly when and where it is needed. For these reasons, inventories are maintained. Inventory control is the technique of maintaining spares and materials at desired levels. It is essential that an optimal level of spares be maintained, which minimizes the cost of holding the item in stock and costs incurred if the spares are not available. It also provides the information needed to ascertain the availability of the required spares for a maintenance work. If the spares are not available, then the action has to be taken to procure the spares and inform the scheduling department when the needed spares become available. Spare parts provisioning and inventory control techniques are discussed in Chap. 5.

1.5.3 Cost Control

The maintenance cost has many components, which are direct maintenance lost production, equipment degradation, backups, and overmaintenance costs. Maintenance cost control is a function of the maintenance philosophy, operation pattern, type of system, and procedures and standards adopted by the organization. It is a major component in the equipment life cycle.

The control of maintenance cost optimizes all the costs in maintenance, while achieving set organizational objectives such as availability, "quality rate," and other efficiency and effectiveness measures. Cost reduction and control can be used as an edge for competition in providing products and services. The issues related to cost and its control are presented in Chaps. 6 and 9.

1.5.4 Quality Control

In the production process, quality of the output may be considered as *fitness for use* and "quality is doing it right the first time." Quality control is exercised by measuring the attributes of the product or service and comparing the same with the product or service specifications, respectively. Maintenance can also be viewed as a process, and the quality of its output can be controlled.

In the case of maintenance work, "doing it right the first time" is very essential. Quality may be assessed as the percentage of accepted maintenance jobs according to the standard adopted by the organization. High quality is usually assured by checking the critical maintenance jobs or by maintenance supervision. The details of maintenance quality control are presented in Chap. 8.

1.6 Managing for Quality and Training

Managing for quality is a managerial responsibility. Usually, maintenance managers/engineers are not fully aware of the importance of improving maintenance production quality. The key for managing for quality lies first in the awareness of the need to improve and second in selecting appropriate improvement techniques. Chapter 14 focuses on the aspects of continuous improvement.

Craftsmen performing substandard maintenance work must be identified. This can be achieved by keeping track of repeat jobs by a given craftsman. Further analysis can be carried out to locate the cause(s) of such substandard work. The likely reasons are non-availability of special tools, craftsman not possessing the needed skill level or poor supervision, etc. Eradicating such causes and monitoring the maintenance work completed can result in improved maintenance production quality.

A forecast of new technologies/processes to be acquired by the company should be made, and craftsmen should be trained ahead of the arrival of the equipment in question. Training issues are covered in Chap. 10.

1.7 Reliability-Centered Maintenance (RCM)

RCM is a high-level strategy that may result in an optimized preventive maintenance (PM) that focuses on system and equipment functions. RCM follows a well-structured system approach for developing the PM. It has been implemented in several industries that include nuclear, energy, and petrochemical. A whole chapter is devoted to RCM in this edition of the book (Chap. 11).

1.8 Total Productive Maintenance (TPM)

TPM is an approach to managing maintenance by bringing total quality management (TQM) approaches and techniques to maintenance. TPM depends on equipment management and employee empowerment. The backbone of TPM is an effective preventive maintenance program. TPM identifies six losses due to maintenance and aims at eliminating them. It attempts to maximize the overall equipment effectiveness. The elements of TPM and its implementation are presented in Chap. 12.

1.9 Intelligent Maintenance

The developments of information and communication technologies provide enablers for effective and intelligent maintenance. These developments made it possible to monitor equipment, transfer data, share information, analyze data with embedded systems, and coordinate activities over the Web. These capabilities made it possible to go beyond predicative maintenance to intelligent prognostics defined in Lee et al. [7] as the systematic approach that can continuously track equipment health degradation and extrapolate temporal behavior of health indicators to predict risk of unacceptable behavior over time as well as pinpoint exactly which components that are likely to fail. This led to the development of the concept of e-maintenance which is an extended use and advance application of condition-based maintenance (CBM). The book will not be current unless the use and the impact of ICT on maintenance are addressed. The book contains two chapters on these issues. Computerized maintenance management system (CMMS) design and assessment are presented in Chap. 9, and e-maintenance system is addressed in Chap. 13.

1.10 Human Behavior

In carrying out the functions of planning, organizing, and controlling, maintenance managers are concerned with how their actions affect human behavior. They should try to know how the behavior of subordinates can affect management's planning, organizing, and controlling actions. In maintenance decision making, behavior of subordinates should be of interest to management. It should be ensured that the desired level of craftsmen satisfaction is achieved and maintained.

1.11 Maintenance Terms

Maintenance management is emerging as a discipline with its own methods, techniques, models, and terms. This section presents the definition of common terms in this emerging discipline. The definitions in this section are based to a great extent on the British Standard [2], 1984 (BS3811:1984).

Availability	The ability of equipment to successfully perform its required function at a stated instant of time or over a stated period of time
Condition-based maintenance	The preventive maintenance initiated as a result of knowledge of the condition of equipment observed through routine or continuous monitoring
Condition monitoring	The continuous or periodic measurement and interpretation of data to infer the condition of equipment to determine its need for maintenance
Breakdown	Failure resulting in the non-availability of equipment
Corrective maintenance	The maintenance carried out after a failure has occurred and intended to restore equipment to a state in which it can perform its required function
Emergency maintenance	The maintenance which is necessary in order to avoid serious consequences, such as loss of production time and unsafe conditions
Failure	The termination of the ability of equipment to perform its required function
Fault	An unexpected deviation from requirements which require corrective action
Feedback	A report on the success or failure of an action to achieve its desired objectives and which can be used to improve a process
Forced outage	Outage due to the unscheduled stopping of equipment
Inspection	The process of measuring, examining, testing, gauging, or otherwise detecting any deviations from specifications
Maintainability	The ability of equipment, under stated conditions of use, to be retained in, or restored to, a state in which it can perform its required function, when maintenance is performed under stated conditions and using prescribed procedures and resources

(continued)

(continued)

Maintenance	The combination of all technical and associated actions by which equipment or a system is kept or restored to a state in which it can perform its designated functions
Maintenance history	A history record showing repair, spares, etc., used to assist maintenance planning
Maintenance schedule	A comprehensive list of items and the maintenance tasks required, including the intervals at which maintenance should be performed
Planned maintenance	The maintenance organized and carried out with forethought, control, and the use of records to meet a predetermined plan
Overhaul	A comprehensive examination and restoration of equipment, or a major part thereof, to an acceptable condition
Preventive maintenance	The maintenance carried out at predetermined intervals or intended to minimize the probability of failure or the performance degradation of equipment
Refurbishment	Extensive work intended to bring equipment up to acceptable functional conditions, often involving improvements
Repair	To restore an item to an acceptable condition by the renewal, replacement, or replacement of damaged or worn parts
Restoration	Maintenance actions intended to bring back equipment to its original conditions
Running maintenance	Maintenance which can be carried out while the equipment is in service
Scheduled maintenance	The preventive maintenance carried out at a predetermined interval of time or number of operations, mileage, etc.
Shutdown maintenance	Maintenance which can only be carried out when the equipment is out of service
Spares stock	Items which are available for maintenance purposes or for the replacement of defective parts
Work order	A written instruction giving detail of work to be carried out including detail of spares and manpower
Work requisition	A document requesting work to be carried out
Work specification	A document describing the way in which the work is to be carried out. It may define the materials, tools, time standards, and procedures

1.12 Summary

Maintenance managers and engineers are assigned activities to plan, design, engineer organize, schedule, execute, and control. They are likely to encounter many problems and must make many decisions. They can frequently simplify these difficulties by using appropriate models; however, if these models do not represent maintenance as an integrated system, most likely they will result in suboptimal solutions. This book advocates a strategic holistic integrated system approach for maintenance and also presents types of methods, models, and techniques necessary for managing and engineering maintenance systems along with examples and case studies. It provides maintenance decision makers and engineers with a set of tools, techniques, and models to optimize and improve maintenance systems.

Exercises

1. Define the word process, and give examples of three processes.
2. Define a system, and give an example of three systems.
3. Which is more general, a system or a process?
4. Select an organization in your area and identify three of their objectives, and then demonstrate how maintenance can contribute to achieving them.
5. How would you obtain feedback in a maintenance system?
6. List the possible maintenance strategies available for maintenance managers.
7. What is the role of job standards in a maintenance system?
8. Define the term maintenance capacity planning.
9. Draw a process chart for maintenance showing all the elements of the process.
10. How would you measure the quality of maintcnancc work?

References

1. Al-Turki OM (2011) Strategic maintenance management. J Qual Main Eng 18(2):87–94
2. British Standard Institute (1964) Glossary of general terms used in maintenance organization, BS 3811
3. Campbell JD (1995) Optimum strategies in excellence in maintenance management. Productivity Press, Portland
4. Duffuaa SO, Raouf A, Campbell JD (1998) Planning and control of maintenance systems: modeling and analysis, 1st edn. Wiley, New York
5. Kelly A (1984) Maintenance planning and control. Butterworth, London
6. Kelly, A (1989) Maintenance and its management, conference communications, Monks Hill, Surry
7. Lee J, Ni J, Dragan D, Hai Q, Haitao L (2006) Intelligent prognostics and e-maintenance. Comput Ind 57:476–489
8. Martin HH (1994) New developments in maintenance. Moret Ernst and Young Management Consultants, Vanolaan
9. Murthy DNP, Atrens A, Eccleston JA (2002) Strategic maintenance management. J Qual Main Eng 8(4):287–305

10 Niebel BW (1994) Engineering maintenance management, 2nd edn. Marcel Dekker Inc., New York

11. Tsang AHC (2002) Strategic dimensions of maintenance management. J Qual Maint Eng 8(1):7–39
12. Tsang AHC (1998) A strategic approach to managing maintenance performance. J Qual Main Eng 4(2):87–94

Chapter 2
Maintenance Strategic and Capacity Planning

2.1 Introduction

Planning is one of the major and important functions for effective management. It helps in achieving goals and objectives in the most efficient and effective manner. Planning is usually divided into three levels, depending on the objective and the planning horizon. The levels are as follows:

1. long-range planning (covers a period of five or more years);
2. medium-range planning (covers 1 month to 3 years); and
3. short-range planning (daily, weekly, and monthly plans).

Maintenance strategic planning is a long-range planning and is concerned with the determination of maintenance mission, strategic goals, and objectives. The mission and the strategic goals are derived and aligned with organization's mission and goals. Then, maintenance strategies and programs are developed to achieve a set mission and goals. Maintenance capacity planning is medium- to long-range planning for maintenance involves the determination of the maintenance resources that are needed to perform the maintenance load in order to achieve organizational objectives such as availability, reliability, quality rates and delivery dates.

Maintenance forecasting is an essential activity for planning. Maintenance forecasting comprises the estimation and prediction of the maintenance load. The maintenance load drives the whole maintenance system, and it consists of two major categories. The first category is the scheduled and planned maintenance which is formed of (1) routine and preventive maintenance, (2) scheduled overhauls which involve closure or plant shutdown, (3) corrective maintenance that involves determining the causes of repeated breakdown and substandard performance as a result of design malfunction, and (4) scheduled overhaul, repair, or building of equipment which is not covered under item 2. The second category is emergency or breakdown maintenance load. This category depends primarily on the failure pattern, and it is a major source of uncertainty in the planning process. The sum of the

S.O. Duffuaa and A. Raouf, *Planning and Control of Maintenance Systems*,
DOI 10.1007/978-3-319-19803-3_2

maintenance load for the two categories is a random variable, and it is the major factor in determining maintenance capacity.

In this chapter, techniques of maintenance load forecasting, strategic planning, and capacity planning are presented. Section 2.2 covers a brief introduction to forecasting. Section 2.3 describes qualitative forecasting techniques. Section 2.4 presents a host of quantitative forecasting models including moving averages, regression analysis, exponential smoothing, and seasonal forecasting. Section 2.5 covers error analysis and measures for testing forecasting models. Section 2.6 outlines the procedure for maintenance load forecasting. Maintenance strategic planning is outlined in Sect. 2.7, and Sect. 2.8 introduces the problem of capacity planning in maintenance. Deterministic techniques for capacity planning are presented in Sect. 2.9. Section 2.10 outlines stochastic techniques for capacity planning. The chapter is summarized in Sect. 2.11.

2.2 Forecasting Preliminaries

Forecasting techniques can be classified into two approaches: qualitative and quantitative. Qualitative forecasting is based on the expert or engineering experience and judgment. Such techniques include historical analogy, surveys, and the Delphi method. Quantitative techniques are based on mathematical models that are derived from the historical data estimates for future trends. These models are either time series-based data such as moving averages and exponential smoothing or structural such as regression models Montgomery and Johnson [7].

A forecasting model is judged by the following criteria: (1) accuracy, (2) simplicity of calculation, (3) data needed for the model and storage requirements, and (4) flexibility. Accuracy is measured by how accurate the model predicts future values and is judged by the difference between the model forecasts and the actual observed values. In general, high accuracy requirements demand a complex relationship and therefore increase the complexity of computation. Flexibility is the ability to adjust to changes in the conditions. In other words, it is a measure of the robustness of the forecasting model. Important considerations in the selection of the forecasting approach are as follows: (1) the purpose of the forecast, (2) the time horizon for the forecast, and (3) the availability of the data for the particular approach. The following are the steps for developing a quantitative forecasting model.

1. Identify the characteristic/item to be forecasted and understand its nature. Define the purpose of the forecast and its time horizon.
2. Screen and validate available data for errors and outliers. Identify the additional data needed and the methodology for collecting it.
3. Use the available data and graphical techniques to hypothesize appropriate models. The model represents a relationship that describes the historical pattern of the data or a relationship between the dependent and independent variables.

4. Use the major part of the data to estimate the parameters of the models. Keep part of it for testing and validating the model. The estimation can be accomplished by an appropriate statistical method such as the least squares or the maximum likelihood.
5. Test and validate the models and select the most appropriate one. Simulation and error analysis are useful tools for testing, validating, and selecting the most appropriate one.
6. Monitor the selected forecasting process and model to detect out-of-control conditions and find opportunities for improving forecasting performance. Improvement can be made by refining parameter estimation or changing the forecasting model. The cycle of the forecasting process is shown in Fig. 2.1

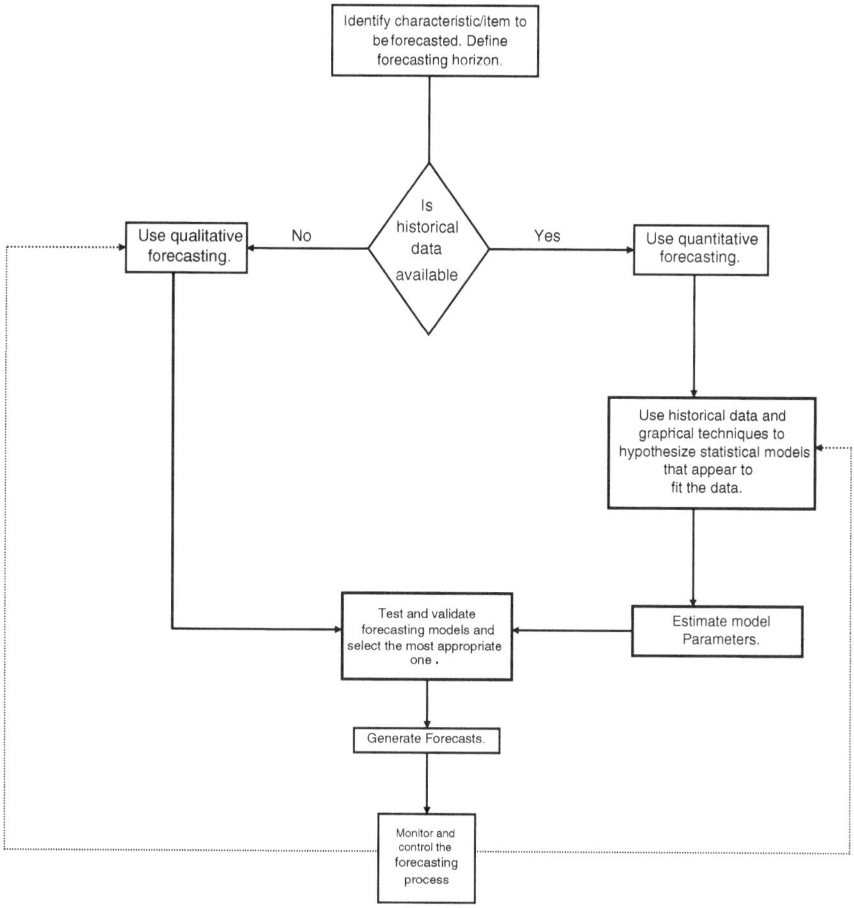

Fig. 2.1 The cycle of the forecasting process

2.3 Qualitative Forecasting Techniques

In the absence of data, the analyst must rely on estimates of experts and their judgment. The role of the analyst in qualitative forecasting is to systematically extract information from the mind of the expert by using structured questionnaires or interviews. He should help the expert or management to quantify their knowledge. Techniques such as cause-and-effect diagrams and the Delphi method can be helpful in identifying relationship among the variables. The analyst should identify which variables influence the forecast and the impact of each one.

After identifying the variables and their impact, the next step is to get an agreement on the magnitude of the variables. Best-case, expected-case, and worst-case scenarios are usually used to estimate the magnitude of the variables. An interactive approach can be used to present arguments to the expert, such as why his estimate differs from the average estimate, and he is asked to revise his/her estimate until a reasonable consensus is reached. When no further reduction in variation about the consensus is possible, the result is used as a forecast.

2.4 Quantitative Forecasting Techniques

In this section, quantitative forecasting techniques are presented. The models presented depend on the availability of the historical data and are usually referred to as time series or structural models. These models either assume future values follow historical trends or that a predictor (independent) variable exists that can provide a model or a functional relationship that predicts the characteristic under study. For example, the age of the equipment can predict the number of maintenance hours required on the equipment. The models presented here include moving averages, regression analysis, exponential smoothing, and seasonal forecasting.

2.4.1 *Simple Moving Average*

Suppose the characteristic under study is generated by a constant process, plus a random error. An example of this could be the load x_t exerted on an electronic component. Mathematically, this can be represented as

$$x_t = b + \epsilon_t$$

where b is a constant and ϵ_t is a random variable with mean 0 and variance σ_ϵ^2.

To forecast the future value, we need to estimate the parameter b. Suppose we have the time series observations, $x_1, x_2, \ldots x_n$. If all the observations are assumed to be equally important, i.e., of equal weight, and if we use the least squares method (see Appendix A, Section A8), then we select a value of b that minimizes the sum of squared error denoted by SS_E.

$$SS_E = \sum_{t=1}^{n} (x_t - b)^2 \tag{2.1}$$

Equation (2.1) is differentiated with respect to b and equated to zero, and Eq. (2.2) is obtained.

$$\frac{dSS_E}{db} = -2\sum_{t=1}^{n} (x_t - b) = 0 \tag{2.2}$$

The estimator of b, $\hat{b}$, is given as

$$\hat{b} = \frac{\sum_{i=1}^{n} x_t}{n} \tag{2.3}$$

which is just the average of the observations on hand. This method generates the next period's forecast by averaging the actual observations for the last n periods.

Example 1 If the maintenance load in man-hours for the last 6 months is given as

Month	1	2	3	4	5	6
Maintenance load	200	300	200	400	500	600

Find the load forecast for periods 7 and 8 using a 3-month moving average.

The forecasted load for month 7 using $n = 3$ is

$$\hat{x}_7 = \frac{400 + 500 + 600}{3} = 500$$

We did not observe x_7, so if the load in month 7 is estimated as 500 as calculated above. The forecast for the 8th month is obtained as follows:

$$\hat{x}_8 = \frac{500 + 600 + 500}{3} = 533.33$$

2.4.2 Weighted Moving Average

It is logical to assume that the most recent observations should have more contributions to future forecasts than to more distant observations, especially when the data are not stable (e.g., there are significant changes). The idea of the weighted moving average is to give each observation a different weight. The forecasting relationship is

$$\overline{x}_{n+1} = \sum_{i=1}^{n} w_t x_t \tag{2.4}$$

where w_i = weight for the ith actual observation

$$\sum_{i=1}^{n} w_i = 1 \tag{2.5}$$

The formula given above can be obtained mathematically by minimizing the sum of the weighted square error using the same procedure as in Sect. 5.4.1. The values of w_i can be determined empirically or be estimated based on experience. If the values of w_i are determined based on experience, this method combines qualitative and quantitative forecasting approaches.

Example 2 Assume, for the data used in example 1, that the most recent period should weigh twice as much as the previous months. Find the forecasts for x_7 and x_8 using three-period moving average:

$$\begin{aligned}
&w_4 + w_5 + w_6 = 1 \\
&w_6 = 2w_5 = 2w_4 = 2w \\
&4w = 1 \\
&w = 1/4, \quad w_5 = w_4 = 0.25 \quad w_6 = 0.5
\end{aligned}$$

Therefore,

$$\hat{x}_7 = 0.25(400) + 0.25(500) + 0.5(600) = 525$$

Assuming the load for the seventh month $x_7 = 525$, the forecast for month 8 is given as

$$\hat{x}_8 = 0.25(500) + 0.25(600) + 0.5(525) = 537.5$$

2.4.3 Regression Analysis

If an independent variable exists that can predict a characteristic (dependent variable) and a reasonable correlation exists between the two variables, then a regression model can be used. For example, if the cost of maintenance for this period, $y(t)$, is a linear function of the number of operational hours in the previous period, $x(t-1)$, then the model would be

$$y(t) = a + bx(t-1) + \epsilon_t \tag{2.6}$$

where σ_ϵ^2.

$y(t)$	desired forecast value for the cost in period t
$x(t-1)$	operational hours in period $t-1$
a, b	parameters to be determined
ϵ_t	a random variable with mean 0 and variance

After estimation, the resulting model that can be used for predicting is of the form:

$$\hat{y}(t) = \hat{a} + \hat{b}x(t-1)$$

It could be possible that the dependent variable is a function of more than one independent variable. In the case of the maintenance cost, it could be a linear function of operational hours in the previous period, $x(t-1)$, and the age of the plant, t. Mathematically, this can be expressed as

$$y(t) = a + bx(t-1) + ct + \epsilon_t \tag{2.7}$$

where a, b, and c are parameters to be determined.

In case of one variable, the parameters a and b can be determined or estimated by finding a good trend line that fits the data points by visual estimation. The parameters a and b are the intercept and the slope of the line, respectively. If more precision is desired, a regression analysis is used. Regression analysis refers to the process of estimating the model parameters using the least squares method. This method fits a line to the observations such that the sum of the square vertical distances from the line is minimized.

The basic equation of a straight line showing a linear trend between an independent variable and a dependent variable $x(t)$ that represents demand for maintenance work is

$$x(t) = a + bt + \epsilon_t \tag{2.8}$$

where a is the intercept and b is the slope that needs to be estimated. The parameter estimation is to determine a and b. The least squares method estimates b and a in terms of $x(t)$ and $t, t = 1, 2, \ldots, n$, as follows:

$$\hat{b} = \frac{n\sum_{t=1}^{n} tx(t) - \left(\sum_{i=1}^{n} t\right)\left(\sum_{t=1}^{n} x(t)\right)}{n\sum_{t=1}^{n} t^2 - \left(\sum_{t=1}^{n} t\right)^2} \tag{2.9}$$

Table 2.1 Data for parameter estimation for the linear regression model

	t	$x(t)$	$tx(t)$	t^2
	1	15	15	1
	2	25	50	4
	3	30	90	9
	4	45	180	16
	5	50	250	25
	6	70	420	36
	7	85	595	49
Summed values	28	320	1600	140

$$\hat{a} = \overline{x} - \hat{b}\overline{t} \tag{2.10}$$

where

$$\overline{x} = \sum_{t=1}^{n} x(t)/n \quad \text{and} \quad \overline{t} = \sum_{t=1}^{n} t/n \tag{2.11}$$

The resulting equation that can be used for prediction is

$$x(t) = \hat{a} + \hat{b}t$$

If the model is taken to be constant plus random variation, i.e., $x(t) = a + \epsilon_t$, then the estimate for a is the average as obtained earlier in Sect. 2.4.1.

Regression analysis can easily be generalized to the case of multiple variables and a polynomial relationship between dependent and independent variables.

Example 3 The monthly maintenance load in man-hours is given in the table below. Develop a straight line that best fits the data and can be used to predict future maintenance load.

Month	t	1	2	3	4	5	6	7
Load	$x(t)$	15	25	30	45	50	70	85

The data set and intermediate computation are given in Table 2.1.

The equation of the straight line is

$$x(t) = a + bt$$

The slope of the line is estimated as follows:

$$\hat{b} = \frac{n\sum_{t=1}^{7} tx(t) - \left(\sum_{t=1}^{7} t\right)\left(\sum_{t=1}^{7} x(t)\right)}{n\sum_{t=1}^{7} t^2 - \left(\sum_{t=1}^{7} t\right)^2} = \frac{7(1600) - (28)(320)}{7(140) - (28)^2} = 11.43$$

$$\hat{a} = \bar{x}(t) - b\bar{t} = \frac{320}{7} - \frac{(11.43)(28)}{7} = -0.005$$

Therefore, $x(t) = -0.005 + 11.43t$.

The above equation can be used for forecasting future load. For example, the load in the ninth month is obtained by substituting 9 in place of t in the equation obtained and is equal to 102.82 man-hours.

2.4.4 Exponential Smoothing

Exponential smoothing is a widely used forecasting method that is simple, efficient, and easy to apply. It assigns weight to observations of previous periods in an inverse proportion to their age. It does this in a very ingenious manner in which only three pieces of data are required to generate the next period's forecast. These are (1) last period's forecast, (2) last period's actual observation, and (3) a smoothing factor, which determines the relative weight given to the recent observation. The basic equation that is the heart of the exponential smoothing is the following:

$$\hat{x}(t) = \alpha x(t-1) + (1-\alpha)\hat{x}(t-1) \tag{2.12}$$

where

$\hat{x}(t)$ the forecast for period t and all future periods in the case of a constant model

$x(t-1)$ actual demand at period $t-1$

$\hat{x}(t-1)$ the forecasted value for $t-1$

α smoothing constant, $0 < \alpha < 1$.

The exponential smoothing approach can be used to estimate the parameters for a constant model, linear model, and any polynomial functional form. The parameter estimation for the constant and the linear cases is given below:

In the constant case, or zero growth, the process model is given as

$$x(t) = b + \epsilon_t \tag{2.13}$$

where

b expected demand in any period

ϵ_t random component having mean 0 and variance σ_ϵ^2.

At the end of the period $t - 1$, we have the observations $x(1),\ldots, x(t - 1)$, from which we need to estimate b and σ_{ϵ}^2. This model is referred to as simple exponential smoothing or

$$x(t) = \alpha x(t-1) + (1-\alpha)\hat{x}(t-1) \tag{2.14}$$

where α is determined using experimentation or judgment. A large value of α (closer to 1) indicates a belief that the current observation carries a high weight. In other words, the system has shifted and the most recent observations resemble its behavior. However, small values of α indicate a belief that the past still resembles to a great extent the system.

If the plotted historical data suggest a linear growth over time, the model is considered a linear model. In that case, the process mean changes linearly with time according to the following equation:

$$x(t) = a + bt + \epsilon_t \tag{2.15}$$

where the expected demand at time t is a linear function of time.

$$E(x(t)|t) = a + bt \tag{2.16}$$

It is known that the lag (the amount by which the forecast deviates from the most recent data value) at the most recent data value in the case of the linear model is

$$\text{lag} = \frac{\beta}{\alpha} \quad \text{slope} = \left(\frac{\beta}{\alpha}\right)b \tag{2.17}$$

If we apply the exponential smoothing again (double smoothing) denoted by

$$\hat{x}(t) = \alpha\hat{x}(t-1) + (1-\alpha)\hat{\hat{x}}(t-1) \tag{2.18}$$

$$\text{lag} = [x(t-1) - \hat{x}(t-1)] = \left[\hat{x}(t-1) - \hat{\hat{x}}(t-1)\right] = \frac{\beta}{\alpha}b \tag{2.19}$$

at each period $t - 1$, the values of a and b are updated as follows:

$$\hat{a}(t-1) = x(t-1) = \hat{x}(t-1) + \text{lag} = 2\widehat{x}(t-1) - \widehat{\widehat{x}}(t-1) \tag{2.20}$$

$$\hat{b}(t-1) = \frac{\alpha}{\beta}\left[\hat{x}(t-1) - \hat{\hat{x}}(t-1)\right] \tag{2.21}$$

Initial conditions to start the process are

$$\hat{b}(1) = \frac{x(t-1) - x(1)}{N-2}$$

$$\hat{a}(1) = x(1)$$

$$\hat{x}(1) = \hat{a}(1) - \hat{b}(1)\frac{\beta}{\alpha}$$

$$\hat{\hat{x}}(1) = \hat{a}(1) - 2\hat{b}(1)\frac{\beta}{\alpha}$$

Example 4 Consider the data given in example 3 for a linear fit by regression. Use exponential smoothing with a linear growth model and $\alpha = 0.2$. Find $\hat{x}(8)$ and $\hat{x}(10)$.

$$\alpha = 0.2$$

$$\beta = 1 - \alpha = 0.8$$

Computing Initial Conditions

$$\hat{a}(1) = x(1) = 15$$

$$\hat{b}(1) = \frac{x(N) \quad x(1)}{N-1} = \frac{85 - 15}{6} = \frac{70}{6} = 11.67$$

$$\hat{x}(1) = 15 - 11.67\left(\frac{0.8}{0.2}\right) = -31.68$$

$$\hat{\hat{x}}(1) = 15 - 2(11.67)\left(\frac{0.8}{0.2}\right) = -78.36$$

The first and double exponential smoothing are given in Table 2.2.

Estimates for a(7) and b(7) can be obtained using Eqs. 2.20, 2.21, respectively.

$$\hat{a}(7) = 2(18) - (-10.67) = 36 + 10.67 = 46.67$$

$$\hat{b}(7) = \frac{0.2}{0.8}[18 - (-10.67)] = \frac{1}{4}[18 + 10.67] = 7.17$$

Table 2.2 The first and double exponential smoothing

t	$x(t)$	$\alpha x(t)$	$\beta\hat{x}(t-1)$	$\hat{x}(t)$	$\alpha\hat{x}(t)$	$\beta\hat{\hat{x}}(t-1)$	$\hat{\hat{x}}(t)$
1	15	–	–	(−31.68)	–	–	(−78.36)
2	25	5.0	−25.34	−20.34	−4.07	−62.69	−66.76
3	30	6	−16.27	−10.27	−2.05	−53.41	−55.46
4	45	9	−8.22	0.78	0.16	−44.37	−44.21
5	50	10	0.62	10.62	2.12	−35.37	−33.25
6	70	14	8.50	22.5	4.5	−26.0	−22.10
7	85	17	19	35	7	−17.68	−10.67

The prediction for any period after L units from 7 is given as

$$\hat{x}(7+L) = \hat{a}(7) + \hat{b}(7)(l)$$
$$\hat{x}(8) = 46.67 + 7.17 = 53.84$$
$$\hat{x}(10) = a(7) + b(7)(3) = 68.18$$

2.4.5 Seasonal Forecasting

In many cases, the process under study might exhibit seasonality characteristics. For example, demand for electricity is high during summer months in the Middle East, or the rate of absenteeism might be higher at the beginning or at the end of the week. Also, the maintenance load might be higher in certain seasons due to weather and operational conditions. The appropriate period to look for seasonality should be driven by the nature of the operation under study. A quick way to check seasonality and growth is by plotting the historical data.

A set of logical steps to be followed when forecasting a characteristic with combined seasonality and growth is as follows:

1. Plot the data and visually determine clear time series characteristics.
2. Determine the growth model and remove the growth component from the data. One way to remove the growth component from the data is to determine an average period for each cycle and divide each data value by the average.
3. Determine whether a significant seasonality is present in the data as it appears with the growth component removed (degrowthed). The seasonality index can be computed by averaging the degrowthed data over the seasons (periods exhibiting the similar behavior).
4. Deseasonalize the original data and analyze the growth factor. A plot of the deseasonalized data will reveal the form of the growth component. The deseasonalizing is accomplished by dividing each data by the appropriate seasonal index.
5. Fit the data by some appropriate method, least squares regression, exponential smoothing, etc.
6. A forecast for the future consists of a combination of seasonal and growth trends.

The above steps are demonstrated on the following data. The following data show the number of hours lost due to absenteeism and late arrivals at the maintenance department for a period of 4 weeks.

The data in Table 2.3 exhibit seasonality. It is clear that the data values for Mondays and Fridays are high. The data for Tuesdays, Wednesdays, and Thursdays are relatively low. The daily averages for each week exhibit a growth model (7, 9, 11, 13). To remove the growth (step 2 above), the data for each day in the week are

Table 2.3 Seasonal data for number of hours lost with five seasons (days) across the period

Week	Days					
	Monday	Tuesday	Wednesday	Thursday	Friday	Average
1	9	3	2	5	16	7
2	11	5	5	9	15	9
3	13	6	7	10	19	11
4	16	8	7	12	22	13

Table 2.4 Seasonal data for number of hours lost with five seasons (days) across the period

Week	Days					
	Monday	Tuesday	Wednesday	Thursday	Friday	Average
1	1.29	0.43	0.29	0.71	2.29	5.01
2	1.22	0.55	0.56	1.0	1.67	5.01
3	1.18	0.55	0.64	0.91	1.73	5.01
4	1.23	0.62	0.54	0.92	1.69	5.0
Total	4.92	2.16	2.03	3.54	7.38	–
Average (I_d)	1.23	0.54	0.51	0.89	1.85	5.02

divided by the daily average (e.g., the Monday value of week 1 is found by dividing $\frac{9}{7} = 1.20$. This has been performed in Table 2.4. The seasonality index for each day is obtained by averaging the degrowthed data over the 4 weeks. Also, Table 2.4 shows the seasonality index for each day in the last row.

The next step is to deseasonalize the data and determine a model for the growth component. One approach for this is to divide each data point by the seasonal index, and then a growth model should be developed for the daily averages. This is shown in Table 2.5.

The daily averages seem to grow linearly. Using linear regression, the parameter for the growth component is given as

$$\hat{x}(t) = 4.15 + 2.36t \tag{2.22}$$

Table 2.5 Deseasonalized data

Week	Days						
	Monday	Tuesday	Wednesday	Thursday	Friday	Daily total	Average
1	7.32	5.56	3.92	5.62	8.65	31.07	6.21
2	8.94	9.26	9.80	10.11	8.11	46.22	9.24
3	10.57	11.11	13.73	11.24	10.27	56.92	11.38
4	13.01	14.81	13.73	13.48	11.89	66.92	13.38

Table 2.6 Weeks 5 and 6 daily forecast with $x(t) = I_d(4.15 + 2.36t)$

Week	Days					
	Monday	Sunday	Wednesday	Thursday	Friday	Daily average
5	19.62	8.61	8.13	14.20	29.51	15.95
6	22.52	9.89	9.34	16.29	33.87	18.31

Table 2.7 Weeks 5 and 6 daily forecast with $x(t) = I_d(5.0 + 2t)$

Week	Days					
	Monday	Tuesday	Wednesday	Thursday	Friday	Daily average
5	18.45	8.1	7.65	13.35	27.75	15
6	20.91	9.18	8.67	15.13	31.45	17

To obtain the daily forecasts for the 5th and 6th week, use the equation above to forecast the weekly average. Then, multiply the weekly average by the daily seasonal index. For example, the forecast for the 5th week average is

$$x(5) = 4.15 + 2.36(5) = 15.95$$

To obtain the forecast for Monday in the 5th week, this value is multiplied by 1.23, the seasonal index for Monday I_m. It is $(1.23)(15.95) = 19.62$. Table 2.6 presents the daily forecast for the 5th and 6th week.

An alternative approach to estimate the growth component is to total or average the data in each cycle. This approach would lead to the daily average for each week and is shown in Table 2.3. These data show a perfect linear trend. Regression analysis for the four daily averages would yield the model

$$\hat{x}(t) = 5.0 + 2t \tag{2.23}$$

Equation (2.23) predicts the daily averages for the 5th and 6th week as 15 and 17. The daily forecast is obtained by multiplying the daily average for the week by the daily seasonality index I_d. The daily forecasts are shown in Table 2.7.

The forecasts given in Tables 2.6 and 2.7 are both logical and sound. The logical questions are which one is better and which one should the decision maker adopt for planning purposes. Error analysis provides an approach for making the selection.

2.5 Error Analysis

Forecast error analysis provides a valid approach for checking the effectiveness of a forecasting model. It also provides a sound methodology for evaluating and selecting from several forecasting models that are available for a particular situation.

The forecast error at period t is the difference between the actual data value $x(t)$ and the forecasted value for it

$$e(t) = x(t) - \hat{x}(t) \tag{2.24}$$

The sum of the errors

$$\sum_{t=1}^{N} e(t) = \sum_{t=1}^{N} [x(t) - \hat{x}(t)] \tag{2.25}$$

is not a valid measure of effectiveness of a forecasting model, but it is a measure of bias. The sum of the errors should approach zero if the model is fitted using the least squares method. The sum of the errors has the problem that large positive errors, $e(t)$, can offset large negative errors. To eliminate this problem, we take either absolute errors or squared errors. The following error measures are commonly used for error analysis and evaluation of forecasting models.

1. Mean absolute deviation (MAD)

$$\text{MAD} = \frac{\sum_{t=1}^{N} |x(t) - \hat{x}(t)|}{N} \tag{2.26}$$

2. Mean-squared error (MSE)

$$\text{MSE} = \frac{\sum_{t=1}^{N} (x(t) - \hat{x}(t))^2}{N} \tag{2.27}$$

3. Mean absolute percent error (MAPE):

$$\text{MAPE} = \frac{100}{N} \sum_{t=1}^{N} \left[\left| \frac{x(t) - \hat{x}(t)}{x(t)} \right| \right] \tag{2.28}$$

4. Mean-squared percent error (MSPE)

$$\text{MSPE} = \frac{100}{N} \sum_{t=1}^{N} \left[\left(\frac{x(t) - \hat{x}(t)}{x(t)} \right)^2 \right] \tag{2.29}$$

One of the above measures can be calculated for all the available models, and the one with the minimum value is selected. This approach can be applied easily for more on some of the models presented in Sect. 2.4 see [3, 7].

2.6 Forecasting Maintenance Work

Prior to performing capacity planning or designing a new maintenance organization, it is essential to have some forecast of the expected maintenance load. The load comprises the following:

1. *Emergency maintenance workload.* This can be forecasted using actual historical workloads and the appropriate techniques of forecasting and/or management experience. This component of the load is random and can be minimized by having a well-designed planned maintenance.
2. *Preventive maintenance workload.* This can be forecasted using actual historical records coupled with newly developed preventive maintenance programs. This should include routine inspection and lubrications.
3. *Deferred corrective maintenance.* This can be forecasted based on historical records and future plans.
4. A forecast for overhaulremoved items and fabrication. This can be estimated from historical records coupled with future plans for improvements.
5. *Shutdown, turnarounds, and design modifications.* This can be forecasted from actual historical records and the future maintenance schedule.

The forecasting of the maintenance load for a new plant is more difficult and must rely on similar plants' experience, benchmarking, management experience, and manufacturers' information (Table 2.8).

Once a plant is in operation, errors in forecasting and job standards may lead to a backlog. An alternative approach is to do forecasting by examining the maintenance backlog. The usefulness of calculating a backlog for planned work is seen when load forecasting is done for the coming week's or month's work. Table 2.8 illustrates this.

The table shows a year-to-date average emergency work level in mechanical trades of 400 h and in electrical, 80 h. This week PM requirements are 1200 h for mechanical and 600 h for electrical. There are an average number of hours required for the minor unplanned work that is done with non-emergency, yet high priority, work orders. Some work was begun last week but was not finished at week's end; it will be completed this week. The total planned priority 3 work in backlog awaiting scheduling—that is work that has been planned and the materials required are

Table 2.8 Weekly backlog report in hours

	Mechanical	Electrical
YTD average "emergency"	400	80
This week's PM level	1200	600
YTD average minor unplanned	600	400
This week's carryover work	800	40
Total planned priority 3 work backlog	12,000	900
Total shutdown backlog	10,000	3000

available—is 12,000 h for the mechanical and 900 h for electrical. Finally, there is a total of 13,000 h of mechanical and electrical work awaiting the next plant shutdown.

This represents the total backlog that is relevant for scheduling the maintenance work for the upcoming week.

2.7 Maintenance Strategic Planning

This section addresses maintenance planning at the strategic level which falls under the long-range planning. In the past, maintenance is not considered as a strategic unit within the organization, but lately this has changed and researchers and maintenance decision makers have brought forward the strategic role of maintenance. Maintenance strategic planning is the process of ensuring the alliance of maintenance mission, goals, objectives, and programs with the organization's mission and objectives. Recent work of Tsang [9], Murthy et al. [8], and Al-Turki [1] has addressed and identified several important issues that are essential in deciding maintenance strategic plan. These issues include maintenance mode of delivery, organization, methodology, and support. The following ten steps are suggested to put to work the frameworks suggested in the recent literature:

- Top management develops organization mission, strategic goals, and objectives.
- Maintenance management (MM) reviews the organization mission and objectives and identifies the role of maintenance in achieving them.
- MM analyzes the current internal and external situations related to the maintenance function using the methodology for determining strengths, weaknesses, opportunities, and threats known as SWOT analysis. SWOT leads to the identification of the strategic issues that need to be addressed. The strategic issues are expected to include mode of delivery, maintenance organization, methodology, manning, and training.
- MM formulates the identified role as a mission statement for maintenance and develops strategic goals and objectives to address the strategic issues.
- Align and prioritize maintenance strategic goals, objectives, and programs with maintenance mission.
- MM checks the alignment of maintenance strategic goals and objectives with organization set mission and strategic goals and objectives.
- MM develops maintenance programs to achieve set strategic goals and objectives.
- Develop a set of quantitative performance measures for the maintenance strategic goals and objectives.
- Evaluate periodically the progress toward achieving the goals and the objectives using the performance measures and identify the gap between the actual and the desired situation.
- Identify the root causes of the gap and implement effective corrective actions.

2.7.1 Maintenance Strategic Issues

In this section, several strategic issues in maintenance that are identified in the literature are discussed. The first issue is the delivery mode. This issue deals with whether maintenance is outsourced, conducted in-house or a combination of both. Each option has its pros and cons. The selection of the right option should be made in light of the set maintenance strategic goals and objectives at the same time to minimize risks and threats to maintenance systems' efficiency and effectiveness. The second maintenance strategic issue is maintenance organization as identified by Tsang [10]. Several options are available in this issue. The options include functional, process, or network. Each of the previous organization options can be crossed with centralized, decentralized, or a cascade option. The third issue is the work structure and control. Many options are available for the work structure depending on the reporting and supervision structure. The fourth strategic maintenance issue is the selection of the support system that includes information system, training, and performance management and reward system. Each element has to be carefully selected to support the overall objective of the organization.

2.8 Maintenance Capacity Planning

Maintenance capacity planning determines the optimal level of resources (crafts, skills, spares, inventory equipment, and tools) to meet the forecasted maintenance load that consists of future load forecast plus the maintenance backlog. An essential element of capacity planning is the determination of the skills of craftsmen, the exact number of various types of craftsmen, the healthy level of backlog, overtime capacity, and contract maintenance. The optimal allocation of the maintenance resources to meet a random and varying workload is a complex and challenging problem. Capacity planning techniques play very important roles in handling this complex problem. The steps involved in capacity planning can be summarized as follows:

1. Determine the total maintenance load.
2. Estimate the required spares and material to meet the load.
3. Determine equipment and tools that are necessary for all types of maintenance work.
4. Determine the skills and the number of crafts from each skill. A special attention should be given to multi-skill crafts.
5. Sometimes in highly computerized equipment, there may be a need to provide a certain specialty that may require special plans.

In maintenance capacity planning, a major issue is to determine the optimal mix of skills of crafts from the available sources to the organization. The usual available sources are regular and overtime in-house crews and contract maintenance. The best mix using these sources is determined using cost and availability measures.

Multi-skilling provides capacity planners with options and alternatives that enrich and generate a wide range of options for determining the optimal capacity. Crafts with multi-skills can be utilized in more than one type of maintenance work. This usually improves manpower utilization.

Techniques for capacity planning can be divided into two major categories: deterministic and stochastic. The deterministic approach assumes that the forecasted maintenance load, standard times, and other random variables are fixed constants. A plan that minimizes cost or maximizes availability depending on the organization objectives that may include reliability, availability, and cost is then determined. In this chapter, two deterministic techniques will be presented, namely:

1. Heuristic tableau method, and
2. linear programming.

The stochastic approach models the maintenance load, standard times, job arrival times, and other variables as random variables with certain probability distributions and uses standard statistical techniques to identify these distributions. Then, a stochastic model is utilized to determine the optimal capacity that meets the maintenance load. Two stochastic techniques will be presented in this chapter, namely:

1. Queuing models and
2. stochastic discrete event simulation

Next, a brief description for these techniques is provided.

2.9 Deterministic Approaches for Capacity Planning

In Sects. 2.9.1 and 2.9.2, two deterministic techniques for capacity planning are presented. These are the heuristic tableau method and linear programming.

2.9.1 Heuristic Tableau Method

In maintenance capacity planning, the heuristic tableau method derives intuitively appealing plans to determine a feasible craft mix, based on sound principles and guidelines. The tableau is used to evaluate the cost of each alternative, and the plan with minimum cost is selected. Sound principles and guidelines include, providing sufficient in-house crafts for high priority work, a reasonable ratio of overtime work to regular time work and a fixed level of a healthy backlog. The following discussion presents the reasoning that may be adopted for determining the required crafts and skills to meet the forecasted mechanical maintenance load using this approach. The same approach can be repeated for other types of maintenance load such as electrical or instrument. As an example, the mechanical workload is first classified into two grades. The skill the work requires and its impact on the facility determines the grade. In this example, it is classified as grade one and grade two mechanical workload, although other classifications based on skill and priority are possible. Prior to presenting the plan in Table 2.9, the following notation is defined as follows:

FM_t	Total forecasted mechanical load
B_{t-1}	Mechanical workload backlog from period $t-1$
TM_t	Total mechanical workload, for period t. $TM_t = FM_t + B_{t-1}$
$TM_{i,t}$	Total mechanical workload of grade i in period t, $i = 1{,}2$
$RM_{i,t}$	Regular in-house capacity for mechanical workload of grade i in period t
$OM_{i,t}$	Overtime capacity for mechanical workload of grade i in period t
$CM_{i,t}$	Contract capacity for mechanical workload of grade i in period t

The plan is derived based on the following commonsense principles and guidelines:

1. All priority work is met by regular in-house crafts as much as possible.
2. If it is not possible to satisfy priority one work by regular in-house crafts use overtime.
3. No backlog is allowed for grade 1 work.
4. The manning level must be determined based on the average maintenance load plus a healthy backlog from grade 2 work.
5. The priority two work is met with overtime or contract maintenance.
6. The overtime capacity is at most 25 % of the regular in-house capacity.
7. In the example, the maximum for the backlog is 100 man-hours. If the backlog exceeds this limit, subcontracting is utilized, and it is assumed that subcontracting can provide as much capacity as needed.

If the periods in Table 2.9 are taken to be weeks and the guidelines (1–6) are taken into consideration, a flexible plan for meeting the maintenance load from the available sources is shown in Table 2.9. Column two contains the forecasted mechanical maintenance load, and column three has the expected backlog from the previous period. Column four contains the total load. Columns five and six present the two grades of maintenance work. The sum of these columns gives column four. Columns seven and eight present the amount of work that will be met by regular in-house maintenance crafts. Columns nine and ten contain the work that is expected to be performed by overtime in-house maintenance. Columns eleven and twelve present the amount of contract maintenance for both grades of mechanical work. The hours in all the columns are standard hours.

It should be noted that this is a target plan; the realized workload might be different from the forecasted load, and an adjustment to the plan at the execution stage might be needed. Different plans can be generated based on the same guidelines, and the one with minimum cost is selected. The required number of employees can be based on the number of standard hours in the regular in-house column, after making an allowance for the expected productivity of the trades.

Table 2.9 Sample data for heuristic tableau capacity planning

Col. #	2	3	4	5	6	7	8	9	10	11	12
Period t	FM_t	B_{t-1}	TM_t	$TM_{1,t}$	$TM_{2,t}$	$RM_{1,t}$	$RM_{2,t}$	$OM_{1,t}$	$OM_{2,t}$	$CM_{1,t}$	$CM_{2,t}$
1	150	100	250	150	100	150	10	–	40	–	–
2	200	50	250	150	100	150	10	–	40	–	–
3	250	50	300	150	150	150	10	–	40	–	–
4	180	100	280	180	100	160	–	20	20	–	–
5	200	80	280	200	80	160	–	40	–	–	–
6	150	80	230	120	110	120	40	–	40	–	–
7	200	30	230	150	80	150	10	–	40	–	–
8	250	30	280	150	130	150	10	–	40	–	–
9	200	80	280	150	130	150	10	–	40	–	–
10	300	80	380	200	180	160	–	40		10	110
11	250	70	320	210	110	160	–	40	–	10	–
12	200	60	260	200	60	160	–	40	–	–	–

The above heuristic tableau approach can be formalized by using a tableau similar to the one used in production planning with some modifications. In this tableau, the sources of manpower supply are shown on the left side of the tableau and the capacity of each source on the right-hand side on the same row. The maintenance load for each period is shown in the columns of the tableau. The cost of meeting the load from a particular source is shown in the corner of each cell in the table.

In production planning products, from the current period can be kept in inventory to satisfy future periods demand, however, this is not the case in maintenance planning. In maintenance, unfinished work is backlogged and performed in future periods at an added cost to the system. It is also possible to divide the maintenance load in each period by skill or priority, and the same method can be applied. Table 2.10 shows a tableau for a three-period maintenance plan for one kind of maintenance work which is the mechanical workload. The same table can be replicated for other types of work. The notations on the tableau are as follows:

C_r hourly cost of mechanical trade on regular time
C_o hourly cost of mechanical trade on overtime
C_s hourly cost of subcontracting
B_t backlog in man-hours at the beginning of period t
CR_t capacity of in-house regular time in period t
CO_t capacity of in-house overtime in period t
CS_t capacity of subcontracting in period t
FM_t forecasted maintenance load in period t

Table 2.10 Data for capacity allocation problem

PERIODS	Period / Sources	1	2	3	Capacity
1	Regular Time	C_r	∞	∞	CR_1
	Overtime	C_o	∞	∞	CO_1
	Subcontract	C_s	∞	∞	CS_1
2	Regular Time	$C_r + \pi$	C_r	∞	CR_2
	Overtime	$C_o + \pi$	C_o	∞	CO_2
	Subcontract	$C_s + \pi$	C_s	∞	CS_2
3	Regular Time	$C_r + 2\pi$	$C_r + \pi$	C_r	CR_3
	Overtime	$C_o + 2\pi$	$C_o + \pi$	C_o	CO_3
	Subcontract	$C_S + 2\pi$	$C_s + 2\pi$	C_s	CS_3
	Maintenance load	M_1	M_2	M_3	

If a work is backlogged for r period and performed at period $r+1$ with regular in-house manpower, it will have a cost of $C_r + r\pi$ per hour. Table 2.10 shows the data needed for a three-period planning horizon. The table shows the costs, capacities, and maintenance load. The symbol ∞ in the cost cell means that a work cannot be done in this period. For example, work that came in period 2 cannot be done in period 1.

A simple least cost heuristic method can be used to compute the allocation of the maintenance load to different sources of manpower supply. The method starts with the least cost cell and satisfies the load as much as possible and then moves to the next least cost until all the load is met. It is highly likely to attain near-optimal solutions with the least cost method. Next, an example is given to demonstrate the tableau approach.

Assume that we have three time periods with maintenance loads 400, 300, and 500, respectively. The in-house regular capacity is 200, 350, and 300 for periods 1, 2, and 3 respectively. The overtime is at most 25 % of the in-house capacity. Subcontracting is abundant, and technically, there is no limit on this source. The cost of performing one in-house man-hour is taken to be 1 unit, overtime man-hour costs 50 % more than this regular time, i.e., 1.5 units, and the subcontracting costs 2 units. Backlogging of one man-hour costs $\pi = 0.3$. The capacity for subcontracting can be taken a large number in this example, and for purpose of demonstration, it is taken to be 500 man-hours. The data and the solution for this situation are shown in Table 2.11.

Table 2.11 Data and solution of the example

Periods	Sources	Period 1 cost	Period 1 allocation	Period 2 cost	Period 2 allocation	Period 3 cost	Period 3 allocation	Capacity
1	Regular Time	1	200	8		∞		200
	Overtime	1.5	40	∞		∞		40
	Subcontract	2	30	∞		∞		500
2	Regular Time	1.3	50	1	300	∞		350
	Overtime	1.8	80	1.5		∞		80
	Subcontract	2.3		2		∞		500
3	Regular Time	1.6		1.3		1	300	300
	Overtime	2.1		1.8		1.5	75	75
	Subcontract	2.6		2.3		2	125	500
Maintenance load		400		300		500		

Applying the least cost heuristic, the plan in Table 2.11 is obtained. The plan calls for meeting the maintenance load by 200 h of regular in-house, 40 h of overtime, and 30 h of subcontracting in the first period. In addition to 50 h of regular time, and 80 h of overtime in the second period. The number of hours backlogged from the first period to the second period is 130. The second-period load is met by 300 h of regular in-house maintenance in the same period. The third-period load is met by 300 h of regular in-house, 75 h of overtime, and 125 h of subcontracting. The total cost (TC) of the plan is sum of the costs in each cell, which is obtained by multiplying the hours in the cell by the cost of the hour.

$$\begin{aligned} \mathrm{TC} = {} & 1 * 2000 + 1.5 * 40 + 2 * 30 + 1.3 * 50 \\ & + 1.8 * 80 + 1 * 300 + 1 * 300 + 1.5 * 75 + 2 * 125 = 1491.5 \text{ units.} \end{aligned}$$

2.9.2 *Linear and Integer Programming for Maintenance Capacity Planning*

Linear programming is a mathematical model that optimizes a linear function subject to linear inequalities. The linear programming model determines the optimal values of decision variables that optimize a given objective such as minimizing cost or maximizing profit. Decision variables are elements under the control of the decision maker, and their values determine the solution of the model. The objective function in linear programming is the criteria with which feasible solutions are evaluated. A feasible solution is a solution that satisfies all the constraints in the system. A constraint is a condition on the system that must be satisfied.

In the case of maintenance capacity planning (MCP), the decision variables could be the number of hours from different skills and trades made available for maintenance capacity through regular in-house, overtime, or contract maintenance. The objective could be to maximize resource utilization or minimize total cost. An example of a constraint is the ratio of overtime hours to regular in-house hours should not exceed certain percentage. In this section, linear programming is introduced via a simple hypothetical example.

Suppose it is forecasted that the mechanical maintenance load is 100 man-hours divided into two grades: 60 h of grade 1 and 40 h of grade 2. Grade classification is based on the skill the work requires and the impact the work has on the facility. The capacity in man-hours for this work can be provided from two skills of workers, i.e., skill 1 and skill 2. Both skills of workers can perform the two types of mechanical work but with different productivity. The productivity of skill one worker is 0.75 and 0.8 for grade 1 and grade 2 mechanical works, respectively. The productivity of skill two worker is 0.5 for grade 1 and 0.7 for grade 2 works. The cost of one man-hour of skill 1 worker is 30 dollars, and the cost of one man-hour of skill 2 worker is 20 dollars. The objective is to determine the number of man-hours from each skill level workers assigned to perform the maintenance load (work) at a minimum cost.

Prior to stating the model constraints and objective, let x_{ij} be the number of man-hours of skill i assigned to perform the mechanical work of grade j, $i = 1, 2, j = 1, 2$. The first constraint is to meet the load for grade 1 mechanical work. This is stated mathematically as

$$0.75x_{11} + 0.5x_{21} \geq 60 \tag{2.30}$$

The second constraint is to meet the grade 2 mechanical work, mathematically stated as

$$0.8x_{12} + 0.7x_{22} \geq 40 \tag{2.31}$$

The non-negativity restrictions state that the man-hours obtained from each skill of workers should be nonnegative, mathematically stated as

$$x_{11}, x_{12}, x_{21}, x_{22} \geq 0$$

The objective function is to minimize the sum of the cost of meeting the required load from the two skill levels. The linear programming model for determining the capacity for the mechanical work in the example above is

$$\text{Min } 30(x_{11} + x_{12}) + 20(x_{21} + x_{22}) \tag{2.32}$$

Subject to

$$\begin{aligned} 0.75x_{11} + 0.5x_{21} &\geq 60 \\ 0.8x_{12} + 0.7x_{22} &\geq 40 \\ x_{11}, x_{12}, x_{21}, x_{22} &\geq 0 \end{aligned}$$

Solving the above model will determine the allocation of man-hours from different skills to perform the two grades of mechanical work at minimum cost. The solution of the linear program could be fractional values. If we want to obtain the solution in whole hours, an additional restriction must be imposed. That restriction requires the variables to be integers. Adding this restriction transforms the linear program to an integer program. In many real situations rounding the linear programming, solution to the nearest hour would be a good solution.

The above formulation can be employed for maintenance capacity planning (MCP) with minor generalization. The formulation will be given for planning the capacity for the mechanical work. The objective is to determine the number of hours from different skill levels from the mechanical trade made available from different sources to perform different grades of mechanical work. The available sources include regular in-house trades, overtime, and contract maintenance. The same model can be used for determining the number of hours needed from other trades such as electrical, instrument, and civil. The following notations are necessary for stating the linear programming model: of a mechanical trade of skill i, from source j when performing mechanical work of grade k

x_{ijkt} Number of man-hours from the mechanical trade of skill i (i = 1, 2, …, I) from source j (j = 1, 2, …, J) made available to perform the mechanical work of grade k (k = 1, 2, …. K) in period t, t = 1, 2, …, T
C_{ij} Hourly cost of a mechanical trade of skill i from source j
P_{ijk} Productivity
F_{kt} Forecasted mechanical load of grade k in period t
B_{kt} Backlog of work of grade k work in period t
UB_k Upper limit for a healthy backlog for grade k work
LB_k Lower limit for a healthy backlog for grade k work
U_{ijt} Upper limit on the availability of skill i mechanical trade from source j in period t
r_{kt} Cost of backlogging one man-hour of grade k work in period t

The linear programming model for determining the required number of mechanical man-hours of different skills, and sources made available to perform all grades of the mechanical work, consists of an objective function and a set of constraints. The objective function is to minimize manpower and backlog cost. The constraints include the work balance constraints, a reasonable ratio of in-house man-hours and overtime to limits on manpower availability. The model is stated as follows:

$$\text{Min} \sum_i \sum_j C_{ij} \left(\sum_k \sum_t x_{ijkt} \right) + \sum_k \sum_t r_{kt} B_{kt} \tag{2.33}$$

Subject to:
Work balance constraints,

$$\sum_i \sum_j P_{ijk} x_{ijkt} + B_{kt} = F_{kt} + B_{k,t-1} \tag{2.34}$$

Limit on overtime man-hours in terms of in-house man-hours (taken as 25 %)

$$x_{i2kt} - 0.25 x_{i1kt} \leq 0 \tag{2.35}$$

Limits on man-hours availability,

$$x_{ijkt} \leq U_{ijt} \tag{2.36}$$

Lower and upper limits on backlog on different work grades,

$$LB_k \leq B_k \leq UB_k$$

$i = 1, 2, \ldots, I$ (number of skills), $j = 1, 2, \ldots, J$ (number of sources), $k = 1, 2, \ldots, k$ (number of grades), $t = 1 \ldots T$ (time period in the planning horizon).

The above model can be solved by any linear programming algorithm such as the simplex-based code in the package Linear Interactive and Discrete Optimizer (LINDO) or the one in the package International Mathematical Software Library (IMSL) or the Optimization Software Library (OSL).

The output of the linear programming model will determine the optimal number of man-hours of different skill levels from a specific trade (mechanical) made available to perform the maintenance work. Based on the number of man-hours from source 1 (in-house), the manning level will be determined.

The linear programming formulation assumed that the parameters of MCP problem are fixed constants, which in reality is not true. In order to examine the sensitivity of the solution obtained, the standard linear programming sensitivity analysis must be conducted. Sensitivity analysis addresses issues, such as what will happen to the solution if the demand changed or a new cheaper resource became available. Also, the complete linear programming solution that includes the dual variables provides information about which resource to increase its level if the demand for maintenance work increased. The interested reader is referred to Taha [9] for more detail on the subject of dual variables and sensitivity analysis.

If we require the crafts and skill levels to be specified in terms of employees to be hired in a maintenance department, we either round up the linear programming solution to the nearest number of employees or reformulate the problem as an integer programming model. This is done as follows:

Let employee of skill *i*, from source *j* in period *t*

n_{ijkt} Number of employees from the mechanical trade of skill i, from source j made available to perform the mechanical load of grade k in period t

$\text{NR} \times S_j$ Number of hours worked per period for an employee from source j (S_j is a constant, and it is usually taken to be 1 if an employee is full time in-house, and between 0 and 1 if an employee is part time or overtime. NH is the number of hours worked by a regular in-house employee per period)

U_{ijt} Upper bound on the availability

C_{ijt} Salary of an employee with skill *i*, from source *j* in period *t*

Assuming that the periods are in months and NH = 160, the integer programming model for determining the number of employees from the mechanical trade of different skills to perform the mechanical load is stated as follows:

$$\text{Min} \sum_i \sum_j \sum_k \sum_t C_{ijt}\, n_{ijkt} + \sum_k \sum_t r_{kt} B_{kt} \tag{2.37}$$

Subject to:

$$\sum_i \sum_j (160 S_j) P_{ijk} n_{ijkt} + B_{Rt} = F_{k,t} + B_{k,t-1} \tag{2.38}$$

$$n_{i2k} - 0.25 n_{i1k} \leq 0 \tag{2.39}$$

$$n_{ijkt} \leq U_{ijt} \tag{2.40}$$

$$LB_R \leq B_{k,t} UB_k \tag{2.41}$$

$$n_{ijk}, \text{integer}$$

The packages mentioned in this section have the capabilities of solving the integer programming model.

2.10 Stochastic Techniques for Capacity Planning

Queuing models and stochastic simulation are two important techniques for capacity planning. Queuing models address the situation where customers arrive at a service facility, perhaps wait in a queue, and then are served by a server and thus leave the facility. In maintenance capacity planning, the customers may take the form of a maintenance job arriving at the maintenance planning and scheduling unit, that job is then planned and routed to a workshop for repair (which represents the service facility). The results from queuing theory allow us to evaluate the performance of such systems under different configurations. Queuing theory has been used to determine maintenance staffing. However, when queuing models are used for capacity planning, the measures of performance are obtained for the system under steady-state conditions which do not represent the transient system behavior which is more close to day-to-day operations.

Stochastic simulation offers a viable alternative when the decision maker is interested in a transient situation or when the system under consideration is complex (which is usually the case in maintenance systems). In stochastic simulation, the maintenance system is represented on the computer and well-designed experiments (scenarios) are used for system performance evaluation. The experiment described for capacity planning of a maintenance system is an allocation of maintenance crews and skills under certain maintenance policy and procedures. The performance measures can be cost, utilization of resources, and availability of critical and major equipment.

In this section, these two stochastic approaches for capacity planning are presented.

2.10.1 Queuing Models

A queuing model can be described as follows: If the customers arrive at a facility, they join a working queue (line). A server chooses a customer according to a certain discipline from the queue and serves him. Upon service completion, the customer leaves the service facility. A queuing model can represent many situations in real life. A relevant example is the arrival of maintenance jobs to a workshop. Other examples include the arrival of customers to a bank and the arrival of telephone calls to a telephone number. The major components of a queuing system are the customers and the servers. The interaction of the customer with the server is gauged by the time the server spends serving the customer. The customers drive the system, and we are interested in their arrival pattern or in their intcrarrival time, which is usually modeled as a random variable. The service time is also modeled as a random variable with a certain probability distribution. Other important elements in a queuing model are: (1) the service discipline, which refers to the manner in which customers are selected for service, e.g., it could be first-come first-serve (FCFS) or some other certain priority rule; (2) the design of the facility, which refers to the number of servers and design of the queues (parallel, series, tandem, network); (3) the queue size (finite, infinite); (4) the size of the source of arrivals (finite, infinite); and (5) human behavior (jockey, balk). In summary, the elements of a queuing model are the following:

(1) arrival distribution,
(2) service time distribution,
(3) design of service facility,
(4) service discipline,
(5) customer population, and
(6) human behavior.

Specifying the above factors will result in a specific queuing model. Some queuing models have steady-state results that give the expected number of customers in the system, L_s, the expected queue length, L_q, the expected waiting time in the system, W_s, and the expected waiting time in the queue, W_q. The above results can be used to evaluate a queuing system.

A notation that summarizes the main characteristic of a parallel queuing model is $(a/b/c)$, $(d/e/f)$ where

a Arrival distribution
b Service time distribution
c Number of parallel servers

d Service discipline
e Maximum number allowed in the system
f Size of customer population

One of the simplest queuing models is the one denoted by (M/M/C) (GD/∞/∞), in which there are *C* servers and interarrival and service times are exponential. If all the servers have exponential service time distribution with parameter μ and if we let $\rho = \lambda/\mu$, then we have the following steady-state results:

$$P_0 = \left\{ \sum_{n=0}^{c-1} \frac{\rho^n}{n!} + \frac{\rho^c}{c!(1-\rho/c)} \right\}^{-1} \tag{2.42}$$

where P_0 is the probability of zero customers in the queue.

$$L_q = \frac{\rho^{c+1}}{(c-1)!(c-\rho)^2} P_0 \tag{2.43}$$

$$L_s = L_q + \rho \tag{2.44}$$

$$W_q = \frac{L_q}{\lambda} \tag{2.55}$$

$$W_s = W_q + \frac{1}{\mu} \tag{2.46}$$

If the above model is used as an approximation for a maintenance department, in which we want to determine the optimal number of repairmen (servers). The steady-state results can be used to evaluate machines' availability and utilization.

Another queuing model known as the machine servicing model is one in which we have *R* repairmen servicing a total of *K* machines, and, because, a broken machine cannot generate a breakdown while in service, the arrival population is finite. The number of repairmen *R* is less than *K*. The objective is to find a value *R* that minimizes the total expected costs that consist of the cost of failure and cost of service. This type of model is denoted by (M/M/R):(GD/K/K), $R < K$.

The model is a special case of the general queuing model. If λ is the rate of breakdown per machine, then if there are *n* broken machines, the arrival rate from the system is given as:

$$\begin{aligned} \lambda_n &= (K-n)\lambda \quad && 0 \le n \le K \\ &= 0 && n \ge K \end{aligned} \tag{2.47}$$

and the service rate is as follows:

$$\mu_n \begin{cases} n\mu & 0 \leq n \leq R \\ R\mu & R \leq n \leq K \\ 0 & n > K \end{cases} \tag{2.48}$$

P_n is the probability of n machines in the system. The steady-state results for the system are derived as follows:

$$P_n = \begin{cases} \binom{K}{n} \rho^n P_0 & 0 \leq n \leq P_0 \\ \binom{K}{n} \frac{n! \rho^n}{R! Rn-R} & R \leq n \leq K \end{cases} \tag{2.49}$$

$$P_0 = \sum_{n=0}^{R} \binom{K}{n} \rho^n + \sum_{n=R+1}^{k} \binom{K}{n} \frac{n! \rho^n}{R! R^{n-R}} \tag{2.50}$$

$$\begin{aligned} L_q &= \sum_{n=R+1}^{K} (n-R) P_n \\ L_s &= L_q + (R - \overline{R}) = L_q + \frac{\lambda_{eff}}{\mu} \end{aligned} \tag{2.51}$$

$\overline{R}$ = expected number of idle repairmen = $\sum_{\mathrm{n}=0}^{\mathrm{R}} (\{R-n\}) P_n$

$$\lambda_{\mathrm{eff}} = \mu = (R - \overline{R}) = \lambda(K - L_s) \tag{2.52}$$

If $R = 1$, the model yields results for a system with single server. If a maintenance system can be represented by such a queuing model, its effectiveness can be evaluated using steady-state measures.

2.10.2 Stochastic Simulation

Maintenance systems have several characteristics that make capacity planning a rather complex problem. These characteristics are as follows:

- Maintenance as a function interacts with other technical and engineering functions in a complex fashion.
- The maintenance factors are highly dependent on each other.
- Maintenance as a function has many uncertain elements. These elements include demand for maintenance, time of arrival of job requests, content, time to complete a job, tools, equipment, and spare parts availability.

- The complexity of the maintenance capacity planning suggests that simulation is one of the most desirable approaches for modeling it.

Stochastic simulation is the process of representing a system on the computer and then employing well-designed experiments (scenarios), to evaluate the system performance. Using this process, systems can be analyzed, planned, and designed. Law and Kelton [6] provide ten major steps for conducting a typical simulation study. In this section, these ten steps are summarized in eight steps and the relationships among them are outlined:

1. *Purpose of simulation*: The first step toward a successful simulation study is to state precisely the purpose of the study. Simulation has been used in maintenance systems for the following purposes: to determine the optimal crew size and staffing, to evaluate the effect of maintenance policies on production systems, to design and plan maintenance operations, and to determine the shutdown time periods.
2. *Simulation models*: The conceptual model used in building the computer simulation study will affect the simulation accuracy and efficiency. The simulation model should contain only the necessary information that captures the essence of the system under study.
3. *Model assumptions*: The assumptions of a simulation model will affect the realism of the simulation results. They also may affect the way results are interpreted. Therefore, each assumption should be reviewed carefully before putting it into effect. Availability of manpower, equipment, job standards, and spare parts are some of the assumptions used in maintenance systems.
4. *Data Accuracy:* Accurate data and their distributions are very essential for a reliable simulation model. To simulate a maintenance system, the distribution of equipment failures and repair times must be identified using sound statistical methodology.
5. *Simulation languages and computers*: One of the major jobs in building a simulation model is to convert the conceptual model into an actual computer simulation program. There are over 100 simulation software programs currently available for a variety of computers. Computer languages such as GPSS, SLAM II, SIMAN, and SIMSCRIPT II.5 are generally used in simulation. Several other criteria also have been used in practice to classify the simulation software.
6. *Program verification and model validation*: Verification is testing and checking the computer code to show that it performs as intended. Validation is to ensure that the model's assumptions are realistic and correct, and the simulation model fairly represents the behavior of the modeled system. Even though this step is fairly tedious and time-consuming, it is the most important step in simulation studies.
7. *Design of experiment*: Another significant element in any simulation study is the design of the experiment. This comprises the following:

   (a) selecting experimental factors,
   (b) selecting measures of performance,

(c) determining the initial conditions,
(d) determining the steady-state conditions,
(e) determining the length of a simulation run,
(f) determining the number of replications, and
(g) applying variance reduction techniques

8. *Output analysis*: In any simulation study, it pays very well to spend time on output analysis. To check for the true estimate, test, validate, and decide on the output results from your simulation, statistical techniques that ensure reliable estimates for system performance must be used. These include deciding on the length of the simulation run, the number of runs, and confidence intervals for estimated measures of performance.

A machine servicing model with general input distributions and general servicing distributions can be used to model maintenance operations and capacity planning. Depending on the maintenance activities and the maintenance load forecasted, the system can be divided into types of jobs requiring different crafts. Craft types include mechanical (ME), electrical (EE), instrumental (INS), civil (CI), and janitorial (JAN). The historical data can be used to develop a probability distribution that represents the arrival of maintenance jobs and the time to complete repairs for a job. The machine servicing model can be modified so that instead of considering physical machines, each type of work generated by the system may be considered as being generated by a hypothetical machine. Each type of work (machine) requires a specific craft or skills. The number of hypothetical machines is determined by the types of work generated by the system. Such a representation can be used within a simulation model to determine the size of crews needed. Figure 2.2 is a schematic representation of a model that can be simulated to determine the staffing levels in a maintenance department.

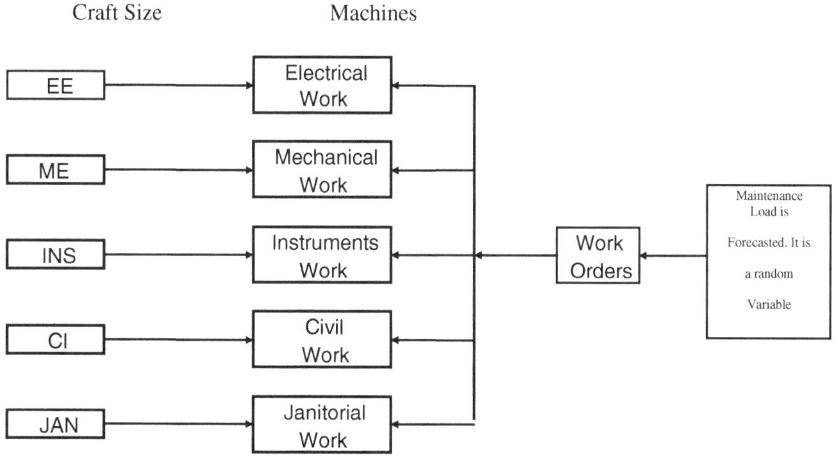

Fig. 2.2 Representation of maintenance capacity planning in a machine servicing framework

In order to implement a simulation model for capacity planning, the following practical steps are needed Duffuaa and Raouf [5].

1. A detailed study of the organization/plant maintenance requirements to determine the types of maintenance crafts and crews required, the types and criticality of equipment repaired, the failure mechanism for each equipment, and the effect of a failure on production or service provided by the organization.
2. Forecast the maintenance workload and divide it according to priority.
3. Outline the existing work order system and define the logic of work assignments (system discipline).
4. Set up the relevant machine servicing model after determining the failure rate of each machine, the service rate, and the cost of each machine being out of service.
5. Develop the simulation software and verify and validate the model.
6. After the model is validated, perform production runs and, on the basis of all measures of performance, find the optimal staffing levels.

2.11 Contract Maintenance

Contracting out maintenance activities is common throughout all industrial and public sectors. The premise is that certain aspects of maintenance can be done as effectively and at a lower cost than in-house resources can accomplish it. A key question is whether contracting maintenance builds or diminishes the competitive advantage of the organization. It is difficult to see how, for example, contracting out custodial services would harm the competitive advantage of a business, unless somehow in-house personnel could do it more effectively and significantly cheaper and with less management attention than a firm specializing in facility cleaning. On the other hand, if haul truck maintenance at a remote mine site is contracted out to a local garage and body shop, it is possible that the mining operation could suffer, regardless of the competitive rates.

In the petrochemical and oil refining sector in North America, it is not uncommon to see the majority of the execution of maintenance contracted to a firm specializing in that business, whereas the reliability engineering and maintenance management is often kept in-house. Major shutdown and overhaul maintenance requires the contracting out of a large segment of the shutdown work backlog, because there is usually a short, finite time period to accomplish all the work and not enough capacity within the organization to accomplish it.

Contracting maintenance during peak periods is more effective in most working environments when finite projects or tasks can be estimated and outsourced as a work package, as one would normally do for a capital expansion or modification project. This allows for a separate contracted crew to work together outside the day-to-day organizational structure of the plant or facility. The work package would specify the contractor's trade skills and numbers and contain any required instructions. Contracting individual trades and integrating them with the in-house staff can lead to inefficiencies and conflicts, as these contracted staff may not know the routines, equipment, or working procedures and rules, thereby reducing the productivity of the entire crew. Further, particularly in a unionized environment, contracting out routine work without extensive bilateral communication on the issue of job security can often lead to labor relations' difficulties and some resentments that may lead to strikes or slowdowns.

2.12 Summary

This chapter presented the essential techniques for maintenance forecasting and strategic and capacity planning. Forecasting is a prerequisite for capacity planning, and capacity planning is a major element in maintenance organization and determines its ability to perform its mission in an effective way. The techniques covered in the forecasting sections of this chapter are moving averages, linear regression, exponential smoothing, and seasonal forecasting. Steps for developing a strategic plan for maintenance are outlined. The methods presented for capacity planning are simple heuristic tableau form, linear programming, queuing theory, and stochastic simulation. The application of the stochastic techniques requires some maturity on the part of the analyst and is introduced only briefly here (for more detail on stochastic techniques see [1–4]).

Exercises

1. The forecasting of the maintenance load that is generated by a newly developed sophisticated equipment does not seem to be amenable to time series forecasting. Why? Suggest a procedure to predict the maintenance load for the next two years for this equipment.
2. In the absence of data, you may have to resort to qualitative forecasting. Explain how would you perform qualitative forecasting and how to validate it?
3. When is it possible to use a control chart to control the qualitative forecasting process? What type of control charts will you use? Explain how.
4. Prove that using the least squares method to estimate the slope, b, and the intercept, a, of the single variable linear fit, $x(t) = a + bt$ results in estimates for b and a as follows:

$$\hat{b} = \frac{n\sum_{t=1}^{n} tx(t) - \left(\sum_{t=1}^{n} t\right)\left(\sum_{t=1}^{n} x(t)\right)}{n\sum_{t=1}^{n} t^2 - \left(\sum_{t=1}^{n} t\right)^2}$$
$$\hat{a} = \bar{x}(t) - b\bar{t}$$

5. Use the least squares method to estimate the parameters of the quadratic model

$$x(t) = a + bt + bt^2$$

6. Derive the simultaneous equations a and b for the model $x(t) = ab^t$ using the least squares method.
7. Given the following data:

t	1	2	3	4	5	6
$x(t)$	5	4	7	8	7	10

Predict $x(t)$ for periods 7 and 8 using five-period moving average and linear regression model. Which resulting prediction will you recommend for use in this case and why?
8. Show quantitatively that the estimate you obtain from the basic exponential smoothing equation is a function of all previous observations; however, the most recent ones are heavily weighted than the far-distant ones.
9. Use a linear model and exponential smoothing to predict the values of $x(t)$ for periods 7, 8, and 10 for the data in Exercise 7.
10. Prove that if a set of data have a noise-free ramp with slope b, the first exponential smoothing will lag the true value by $\left(\frac{\beta}{\alpha}\right)b$ where $\beta = 1 - \alpha$
11. How would you determine α optimally in the exponential smoothing model?
12. Consider the following quarterly data for the maintenance load in man-hours for the last 5 years.

Year	Quarter			
	I	II	III	IV
1	215	120	150	100
2	250	175	75	150
3	300	250	75	165
4	350	275	100	200
5	400	300	125	225

(a) Determine an appropriate seasonal index for each quarter.
(b) Deseasonalize the data and fit it to an appropriate growth model.
(c) Predict the quarter values for the 6th and 7th years.

13. Discuss the advantages and disadvantages of the approaches given in this chapter for capacity planning.
14. Suppose you were asked to develop a capacity plan for a maintenance department. Which approach will you select from the approaches given in this chapter? and why?
15. For the data in Table 2.9, evaluate the cost of the plan given in the table if an in-house regular hour is 10$, overtime hour is 15$, and a contract man-hour costs 25. In addition, every backlog man-hour delayed from period t to $t + 1$ costs 5. Develop an alternative competitive plan in which you can have instead of 5 skilled employees have a mixture of skill and unskilled. The hour of unskilled worker is 8$.
16. Develop a linear programming model for the data given in Table 2.9. Assume that the trades in problem 15 are available.
17. Explain how would you use the machine servicing model for determining optimal mix of crafts and skills to meet the maintenance load.
18. What are the disadvantages of using linear programming for capacity planning.
19. Locate a factory near your area and study its operations and maintenance.
    (a) Forecast their maintenance load.
    (b) Use the structured tableau method to determine their maintenance capacity in terms of staff only.
    (c) Use linear and integer programming to plan their maintenance capacity.
20. Apply stochastic simulation to plan the maintenance capacity for the factory in problem 19. Is there a difference between the approaches in problems 19 and 20. Why do you expect such a difference? Which is more appropriate for this case and why?

References

1. Al-Turki OM (2011) Strategic maintenance management. J Qualit Maint Eng 18(2):87–94
2. Andijani AA, Duffuaa SO (1996) A critical evaluation of simulation study in maintenance. In: Proceedings of the 1st international conference on industrial engineering, theory and applications, Houston, Texas, Dec 1996
3. Bedworth DD, Bailey JE (1987) Integrated production systems and contro, 2nd edn. Wiley, New York
4. Duffuaa SO, Andijani AA (1996) Stochastic simulation: an effective tool for modeling maintenance systems. A working paper, Department of Systems Engineering, KFUPM, Dhahran 31261, Saudi Arabia
5. Duffuaa SO, Raouf A (1995) A simulation model for determining maintenance staffing in an industrial environment. Simulation 59:293–299
6. Law AM, Kelton WD (1993) Simulation modeling and analysis. McGraw Hill, New York
7. Montgomery DC, Johnson LA (1976) Forecasting and time series analysis. McGraw Hill, New York

8. Murthy DN, Atrens P, Eccleston JA (2002) Strategic maintenance management. J Qualit Maint Eng 8(4):287–305
9. Taha H (1992) Operations research: an introduction, 5th edn. Macmillan Publishing Company, New York
10. Tsang AHC (1998) A strategic approach to managing maintenance performance. J Qualit Maint Eng 4(2):87–94

Chapter 3
Preventive Maintenance, Concepts, Modeling, and Analysis

3.1 Introduction

Maintenance work can be divided into two broad categories: planned and unplanned work.

Planned work implies that, all resources necessary to accomplish the tasks have been preplanned and are available and, the work is to be performed according to a set schedule. Unplanned work may have a set of standard instructions available, may have the necessary skills and parts nearby, or may be slotted into a maintenance schedule on an ad hoc basis, but does not meet both the preplanned and prescheduled criteria.

Planned maintenance refers to maintenance work that is performed with advance planning, foresight, control, and records. Preventive maintenance (PM) is a series of preplanned tasks performed to counteract known causes of potential failures of those functions. It is carried out to ensure equipment availability and reliability. Equipment availability can be defined as the probability of equipment being able to operate whenever needed. The equipment reliability is the probability that the equipment will be functioning at time $t + \Delta t$, given that it was functioning at time t.

Preventive maintenance is considered as "planned" work, the distribution by craft hours in a well run industrial maintenance facility is expected to be as described in Fig. 3.1.

Preventive maintenance can be conditioned based or based on the historical data of equipment failure. Figure 3.2 shows a makeup of preventive maintenance. It consists of two categories; these are statistical and reliability based or condition based. The first category is based on data obtained from the equipment historical record. The second category is based on the equipment performance and condition. Next detailed description of preventive maintenance is given in Sect. 3.2.

The purpose of this chapter is to present elements of planned and preventive maintenance concepts and models. Preventive maintenance concepts are presented in Sect. 3.2 followed by a host of diagnostic techniques that are useful for

S.O. Duffuaa and A. Raouf, *Planning and Control of Maintenance Systems*,
DOI 10.1007/978-3-319-19803-3_3

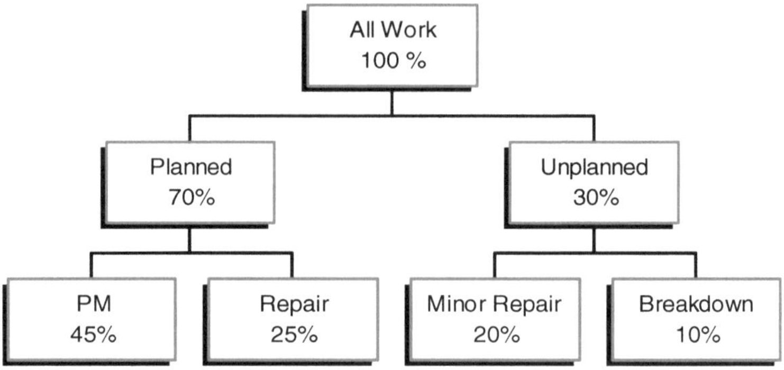

Fig. 3.1 Maintenance work distribution

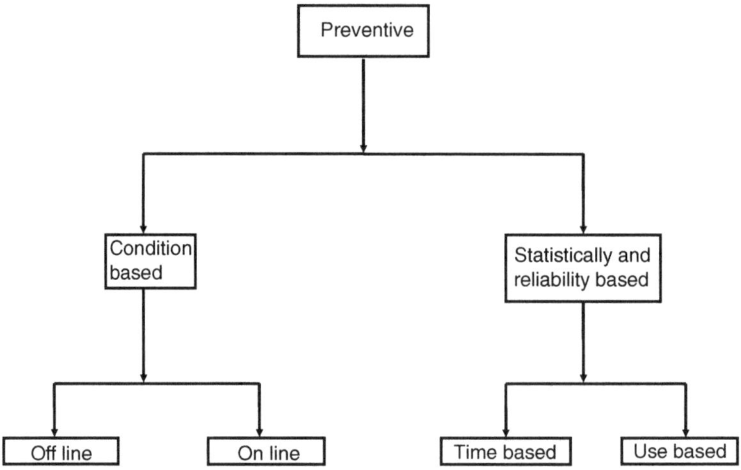

Fig. 3.2 Categories of preventive maintenance

implementing condition-based maintenance (CBM) programs in Sect. 3.3. Section 3.4 describes the concepts that are necessary in replacement decisions. Section 3.5 outlines the steps for developing a planned maintenance program in a simple straightforward manner. Section 3.6 presents a host of mathematical models for optimal preventive maintenance and replacement, and Sect. 3.7 presents several mathematical models for inspection. These models aid in determining the frequency of inspections that will allow determining the condition of equipment in order to plan an appropriate maintenance action. Section 3.8 outlines advanced concepts in modeling preventive maintenance when the maintenance action is imperfect, and Sect. 3.9 outlines the concept of delay time modeling and its use for determining preventive maintenance. Section 3.10 summarizes and concludes the chapter.

3.2 Preventive Maintenance

PM is defined as a series of preplanned tasks performed to counteract known causes of potential failures of the intended functions of an asset. It can be planned and scheduled based on time, use, or equipment condition. It is the preferred approach relative to breakdown maintenance, for four primary reasons:

- The frequency of premature failures can be reduced through proper lubrication, adjustments, cleaning, and inspections triggered by performance measurement.
- If the failure cannot be prevented, periodic inspection and measurement can help to reduce the severity of the failure and its possible domino effect on other components of the equipment and thereby mitigate the negative consequences to safety, the environment, or production throughput.
- Where we can monitor the gradual degradation of a function or parameter, such as product quality or machine vibration, a warning of impending failure may be detected.
- Finally, because an unplanned interruption is often extremely damaging to production schedules and output, and the actual cost of an emergency breakdown is higher than a planned one, and the quality of repair can suffer under the pressure of an emergency, preventive maintenance is preferred to unplanned maintenance.

The most critical question in preventive maintenance is what task or series of tasks should be performed to prevent failure. Clearly, if we understand the actual failure mechanism of the equipment, we can then decide what tasks are logical to prevent failure and which ones are irrelevant. If the PM program requires changing the headlights on an automobile each month, will that have an impact on the failure rate? Most likely not, because the failure mechanism is not related to time, but to other variables, such as the distance, the car is driven behind a large truck on a gravel road.

If the dominant failure mechanism is time based or wear, i.e., if the probability of a failure gradually increases with time, age, or usage, then the maintenance tasks may be time based. If however, the probability of a failure is constant regardless of time, age, or usage, and there is a gradual degradation from the onset of failure, then the maintenance tasks may be condition based. Time-based tasks are warranted if the periodic restoration or component replacement restores the equipment to perform its intended function. This task could range in complexity from a comprehensive overhaul of the entire unit to a simple job such as a filter replacement.

Condition-based tasks, warranted when the failure prevention approach is unknown, focus on measuring a parameter which indicates a deterioration or degradation in the functional performance of the equipment. The measurements and inspections themselves may be regularly scheduled, but the restorative or preventive tasks are not. These measures can be related directly to the machine operations, such as vibration, running temperature, amperage drawn, the lubricating oil contaminants, or noise level, or can be a surrogate measure of machine operation, such as the product quality, its dimensions, wear patterns, or composition. The actual failure distributions, or conditional probability of failure curves, have been developed most extensively for equipment systems and components in the airline industry. These studies suggest that the predominant patterns show a constant probability of failure with age, with the exception of "infant mortality" or failure in the earliest period after commissioning. Therefore, when equipment is new—or recently overhauled—it exhibits a probability of failure higher than at a later period. This is attributed to possible design, manufacture, overhaul or installation errors, or inappropriate initial operating and maintenance procedures. Once these are corrected, failure is virtually unrelated to age.

If we look at rolling bearings as an example, only a small percentage of bearings actually fail in service, but usually outlive their host equipment. Most bearing failures occur because of poor or improper lubrication, solid or liquid contaminants entering the bearing, or because of improper handling or mounting. When bearings are properly handled, mounted, aligned, lubricated, sealed, and kept from extreme temperatures, their predominant failure mode is fatigue—old age. Therefore, non-intrusive condition-based preventive maintenance is the logical choice for monitoring bearings—oil analysis, vibration, temperature, or noise.

Time-based maintenance, e.g., overhauls, is technically feasible if the item has an average identifiable life. Most items survive to that age and the action restores the item's condition to its desired function. CBM is technically feasible if it is possible to detect degraded conditions or performance, there is a practical inspection interval, and the time interval (from the inspection to the functional failure) is long enough to allow corrective actions or repairs. Figures 3.3 and 3.4 illustrate the situations when to use time-based or condition-based maintenance tasks.

Because complex equipment and components will have several plausible failure causes, it is necessary to develop a series of preventive maintenance actions—some condition-based and some time-based for the same equipment—and to consolidate these into a PM program. The program will have tasks grouped by periodicity, i.e.,

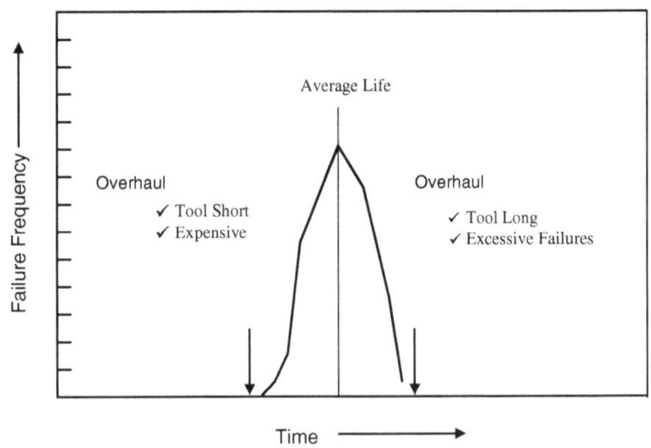

Fig. 3.3 Time-based overhaul

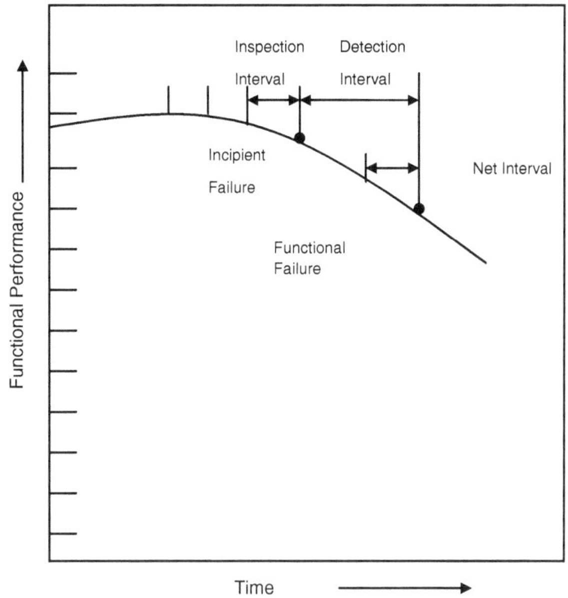

Fig. 3.4 Condition-based maintenance

daily, weekly, annually, operating hours, and cycles, and grouped by trade, i.e., mechanic, electrician, operator, and technician (Table 3.1).

Preventive maintenance is the prime requirement to reduce the frequency and severity of equipment breakdowns. Three broad measures are used to monitor the comprehensiveness of the PM program:

- PM coverage—the percentage of critical equipment for which PM has been developed
- PM compliance—the percentage of PM routines that have been completed according to their schedule
- Work generated from PM routines—the number of maintenance actions that have been requested originating from the PM routines.

Table 3.1 Preventive maintenance program actions

Who	When	Action	Equipment	Conditions	Measure
Mechanic	Weekly	Inspect/ adjust	Hydraulic system	Pressure	2500 ± 50 psi
Technician	Semi-annually	Take reading	Motor bearings	Vibration	Octave band to baseline
Operator	Monthly	Lubricate	Gear motor	Level on dipstick	Top-up to high indicator with 10 W 40 oil
Operator	Daily	Check	Gear motor	Oil pressure	Replace oil filter with P-OF-4201-86 if delta $p > 10$ psig

Condition-based preventive maintenance requires monitoring a variable that is closely related to equipment failure. It is very necessary to identify which parameter is used to monitor and measure. Next, a host of diagnostic techniques is presented that will aid in developing a CBM system.

3.3 Diagnostic Technologies

Before the development of technologies used to assess the condition of equipment, operators and maintenance personnel relied on their own senses: touch (temperature, vibration, wear), smell (temperature, contamination), sight (vibration, temperature, alignment), hearing (noise, vibration, cavitation, wear), taste (contamination) and that "sixth" sense which is duplicated today as expert diagnostic systems. The objective of the inspection was to look for a warning of an impending failure, so that the repair could be planned, scheduled, and completed to minimize the impact on operations and total cost. The key difficulty with using human senses is the subjectivity of the data collection and interpretation, and the amount of time available to react once the condition is determined.

Diagnostic technologies have become widespread in all industrial sectors over the past several decades. The most commonly applied CBM techniques are vibration analysis, oil analysis, thermography, ultrasonic, electrical effects, and penetrants.

3.3.1 Vibration Monitoring and Analysis

Vibration can be defined as the movement of a mass from its point of rest through all positions back to the point of rest, where it is ready to repeat the cycle. The time it takes to do this is its period, and the number of repetitions of this cycle in a given time is its frequency.

The severity of vibration is determined by the amplitude, or maximum movement of its peak velocity and peak acceleration. The phase angle is often measured when comparing the motion of a vibrating part to a fixed reference. Machines will vibrate across a wide spectrum of frequencies. Vibration analysis in condition monitoring is accomplished by comparing vibration characteristics of current operation to a baseline, which was measured when the machinery was known to be operating normally. The selection of the specific parameters to be measured depends primarily on the frequency of the vibration.

Vibration analysis techniques can be used to monitor the performance of mechanical equipment that rotates, reciprocates, or has other dynamic actions. Examples include gearboxes, roller bearings, motors, pumps, fans, turbines, belt or chain drives, compressors, generators, conveyors, reciprocating engines, and indexing machines.

Three common types of vibration analysis are the following:

Broadband vibration analysis monitors the total machine train and is useful for reviewing basic information and trends, but of limited use for pinpointing problem areas.
Octave band vibration analysis is more useful, with the spectrum divided into a series of ranges which can be compared to predetermined values for looking at deviations in vibration frequency.
Narrowband frequency analysis is most useful as a diagnostic tool, with the capability of determining the specific problem areas and causes.

3.3.2 Oil Analysis

When analyzing the oil from a machine, there are a number of different techniques that can be applied to look at the chemical composition of the oil and foreign materials in it.

Ferrography and magnetic chip detection examine the iron-based wear particles in lubrication oils to determine the type and extent of wear and can help determine the specific component that is wearing.

Spectrometric oil analysis measures the presence and amounts of contaminants in the oil through atomic emission or absorption spectrometry. It is useful for determining not only iron, but also other metallic and nonmetallic elements, which can be related to the composition of the various machine components, such as bearings, bushings, and piston rings. It is useful when wear particles are initially being generated in the early stages of failure, as those particles are small.

Chromatography measures the changes in lubricant properties, including viscosity, flash point, pH, water content, and insoluble fraction, through selective absorption and analysis.

3.3.3 Thermography

The most common uses for thermography are to determine poor electrical connections, hot spots, furnace and kiln refractory wear, critical boiler, and turbine component overheating. Thermography uses infrared radiation to measure surface temperature. An infrared camera shows surface temperature variations, calibrated to provide the absolute temperature or temperature gradients through black and white or color variations.

3.3.4 Ultrasonic

There are several techniques for ultrasonic testing, but they all are used to determine faults or anomalies in welds, coatings, piping, tubes, structures, shafts, etc. Cracks, gaps, buildups, erosion, corrosion, and inclusions are discovered by transmitting ultrasonic pulses or waves through the material and assessing the resultant signature to determine the location and severity of the discontinuity. This technique is also used to measure flow rates.

3.3.5 Electrical Effects Monitoring

There are several tests for corrosion using a simple electric circuit monitored by varying degrees of sophisticated instrumentation. The Corrater uses the electrochemical polarization method in a vessel with corrosive liquid. The Corrometer uses the electrical resistance across a probe inserted in the active environment, e.g., refinery process equipment.

The most common devices used for monitoring or testing motors or generators are voltage generators, including mergers. These measure the resistance of insulation and apply a test voltage from 250 to 10,000 V.

3.3.6 Penetrants

Electrostatic and liquid dye penetrants are used to detect cracks and discontinuities on surfaces which are caused in manufacturing by wear, fatigue, maintenance and overhaul procedures, corrosion, or general weathering. The penetrant is applied and allowed to penetrate into the anomalies. The surface is cleaned and the penetrant revealed through direct visual, fluorescent, or electrostatic techniques.

3.4 Replacement Decision Models

The decision to repair, upgrade, or replace equipment or repairable spares can be taken at the design stage for a new system, at a point in the equipment's life cycle where a breakdown has occurred or when obsolescence becomes evident. It requires a look at total life cycle costs of the item for the remaining portion of its life cycle. Total equipment or item's life cycle includes the time from its acquisition to its eventual disposal. All costs for initial purchase, installation, repair, replacements, upgrading, movement to and from repair facilities, removal from service, dismantling, and disposal are considered.

Repairable items have an additional consideration to repair or replace the item when it breaks down. Is the item physically repairable? For complex items such as turbines, the answer is usually "yes" and for many component items such as filters, the answer is often "no." Large single-piece engineered components such as shafts, impellers, rotors, and machine casings are very expensive and repair may prove to be more economical than replacement. If it is known that the original equipment manufacturer no longer makes the item and there is no known substitute, it is often necessary to repair the item at any cost, or to elevate the decision to the level of replacement of the next higher assembly versus repair of the unserviceable item.

The decision to repair should also be accompanied by a decision regarding the most economical repair crew to utilize. The cost and capability of repair using in-house resources, also known as organic repair, should be compared with the cost of shipping the item to and from a contractor or support repair depot for repair.

The best time to make replacement decisions for repairable items is during the design phase of the equipment's life cycle. Much of the information needed to make the decision is readily available from the equipment supplier. Years after, the equipment is acquired that type of information becomes more difficult to obtain because the supplier may no longer be supplying the same equipment. Replacement costs for the equipment and its components can be taken directly from the equipment purchase costs and recommended spares listings. Repair costs can be requested from the supplier and from repair facilities recommended by the supplier for comparison with ones own in-house estimates. Failure rate data will be necessary to predict the frequency of repair events for a total life cycle cost calculation. Again, this will be easier to obtain at the time of equipment purchase, but it may not be as accurate as failure records specific to the equipment application. In the absence of manufacturer failure data, it is possible to use generic data from industrial failure rate and reliability data bases.

The total life cycle cost of an item consists of the sum of acquisition, investment, operation and support, and disposal costs. Each of those has several component costs, but the decision to repair or replace enters primarily into the evaluation of initial investment, operation and support, and sometimes disposal costs.

$$C_{\mathrm{LC}} = C_{\mathrm{Acquision}} + C_{\mathrm{investment}} + C_{\mathrm{O\&S}} + C_{\mathrm{Disposal}} \tag{3.1}$$

where LC is life cycle and O&S operation and support.

Acquisition costs are often fixed and include the design, evaluation, engineering, and project management costs for the item.

Initial investment costs may differ when comparing repairable versus non-repairable designs. In particular, the costs of initial sparing may vary significantly depending on whether the spares are repairable or not. If spares are repairable, the costs of those repairs must be considered as part of the operation and support costs. Support equipment for the troubleshooting, repair, and testing of the item being considered must also be taken into account in the initial investment costs. Support equipment can include anything from special-purpose tooling to computerized diagnostic equipment and programming. Also, the cost of technical documentation required to support repair activities should be considered as it is often more expensive to acquire this detailed documentation than to accept the manufacturer's standard operation and maintenance manuals.

Operation and support costs include the costs of consumable items used in the repair and operation of the item, replacement spares costs, inventory carrying costs, repair personnel costs, repair facilities costs, direct maintenance costs, packaging, handling and storage costs, technical support costs, and the costs of future modifications.

The costs of disposal may also be influenced if competing design options require special handling or disposal arrangements, or waste management provisions.

It is within the operation and support cost category that special attention is required to assess replacement versus repair options. The costs that must be taken into account include the following:

- Repair in situ;
- Removal, discard, and replacement with spares;
- Removal, replacement with spares, repair, and return of repaired item to inventory; and
- Which repair agent carries out the work.

In situ repair means repairing the equipment without the replacement of repairable items with spares. This can often make the repair time and associated labor costs increase and result in less operational time for the system in which the equipment is installed.

The replacement of failed assemblies with spares often requires less time out of service but does require the stocking in inventory of the spares expected to be required for those repairs. Stocking levels must be decided upon in advance, and the stock represents an investment on the part of the owner and generates the need to expend carrying costs so long as the stock is in inventory and taking up space in the plant. The amount of stock to carry can be calculated using various decision models, which are available commercially, and discussed in Chap. 7.

The failure mechanism influences the selection of an appropriate course of action needed to be taken for equipment replacement. For example, truly random failures cannot be predicted or prevented and cannot be eliminated by time-based replacement. Nonetheless, it is necessary to decide on how to handle the random event. If the random failure starts small and progresses to some point where the equipment is rendered unserviceable, it may be possible to use time-based or continuous condition monitoring techniques to identify the incipient failure before it progresses too far and then base a replacement decision on the condition seen. Many random failures follow that pattern and can be handled as described. The economic decision about repair or replacement would be needed to determine the action to take for the failed components (which are removed). Other failures involving wearing mechanisms such as cyclical stress or thermal reversals are totally dependent on time or usage and lend themselves well to replacement without condition monitoring on a timed basis. Again, the decision to repair or replace the worn or expended item is based on an economic analysis as described above.

3.5 Elements of Planned Maintenance

Planned maintenance refers to maintenance work that is performed with advance planning, foresight, control, and records. It includes the whole range of maintenance, and it applies to replacement, preventive, and breakdown strategies. It is characterized by the following:

- The maintenance policy has been stated carefully.
- The application of the policy is planned in advance.
- The work is controlled to confirm the original plan.
- Data are collected, analyzed, and used to provide direction for future maintenance policies.

In this section, the steps for developing a planned maintenance program are presented.

3.5.1 Plan Administration

The first step toward developing a comprehensive planned maintenance program is to put together a task group that initiates and develops the plan. A single person should be appointed to head the task group, and management commitment is essential for the plan. After announcing the plan and forming the necessary organization for it, the task group should embark on putting together the program.

3.5.2 Facility Inventory

The facility inventory is a list of all facilities including all equipment in a site. It is made for the purpose of identification. An inventory file of all equipment should be developed showing equipment identification, description of facility, location, type, and priority (importance).

3.5.3 Identification of Equipment

It is essential to develop a system by which each equipment is identified uniquely. A coding system that helps in this identification process should be established. The code should indicate location, machine type, and machine number. This coding system should differ from plant to plant, and its design should reflect the nature of the facility.

3.5.4 The Facility Register

The facility register is a file (electronic or hard copy) including technical detail about items that are included in the maintenance plan. These data are the first to be fed to the maintenance information system. The equipment (item) record should include an identification number, location, type of equipment, manufacturer, date of manufacturing, serial number, specification, size, capacity, speed, weight, power service requirement, connection details, foundation detail, overall dimension, clearance, reference drawing number, reference number for service manuals, and interchangeability with other units.

3.5.5 Maintenance Schedule

A maintenance schedule must be developed for each equipment in the program. The schedule is a comprehensive list of maintenance tasks to be carried on the equipment. The schedule includes the name and identification number of the equipment, location, reference number of the schedule, detailed list of tasks to be carried out (inspections, preventive maintenance, replacements), the frequency of each task, the crafts needed to carry out the task, time for each task, special tools needed, material needed, and details of any contract maintenance arrangements.

3.5.6 Job Specification

The job specification is a document describing the procedure for each task. It is intended to provide the details of each task in the maintenance schedule. The job specification should indicate the identification number of the item (equipment), the location of the item, the maintenance schedule reference, the job specification reference number, the frequency of the job, crafts required for the job, the details of the task, components to be replaced, special tools and equipment needed, reference drawings, and manuals and safety procedures to be followed.

3.5.7 The Maintenance Program

The maintenance program is a list allocating maintenance tasks to specific time periods. When developing the maintenance program, a great deal of coordination must be done in order to balance the workload and meet production requirements. This is the stage at which the planned maintenance is scheduled for execution.

3.5.8 *Program Control*

The maintenance program developed must be executed as planned. Close monitoring is needed in order to observe any deviation from the schedule. If deviations are observed, a control action is needed. The methods of control are detailed in Chap. 4.

3.6 Mathematical Models for Optimum Preventive Policies

In this section, several maintenance policies for systems that are subject to stochastic failure are defined and mathematical models to determine the optimum level for each policy are formulated.

Two basic preventive maintenance policies proposed by Barlow and Hunter [3] are examined in the literature extensively. These are age-based and constant interval replacement polices known as type I and type II policies. The statements of the polices, their models, and generalizations are given in the next sections. In this part, the notations necessary for the formulation of the models are stated.

C_p	cost of preventive maintenance
C_f	cost of breakdown (failure) maintenance
$f(t)$	time to failure probability density function (p.d.f.)
$F(t)$	equipment or system failure distribution, and it is the integral of $f(t)$
$r(t)$	failure rate function
$N(t_p)$	number of failures in the interval $(0, t_p)$; $N(t_p)$ is a random variable
$H(t_p)$	expected number of failures in the interval $(0, t_p)$
$R(t)$	reliability or survival function
$M(t_p)$	expected value of the truncated distribution with p.d.f. $f(t)$ truncated at t_p
$M(t_p)$	$\int_{-\infty}^{\infty} tf(t)\mathrm{d}t/\left[1 - R(t_p)\right]$
$C(t_p)$	expected cost per cycle
$\mathrm{UC}(t_p)$	expected cost per unit time

3.6.1 *Optimal Age Preventive Replacement (Type I Policy)*

Policy I (age preventive replacement) is defined as follows: perform preventive replacement after t_p hours of continuing operation without failure; t_p could be finite or infinite. In case of an infinite $t_{p,}$ no preventive maintenance (replacement) is scheduled. If the system fails prior to t_p hours having elapsed, perform maintenance (replacement) at the time of failure and reschedule the preventive maintenance after t_p operation hours. In this policy, it is assumed that the system is as good as new

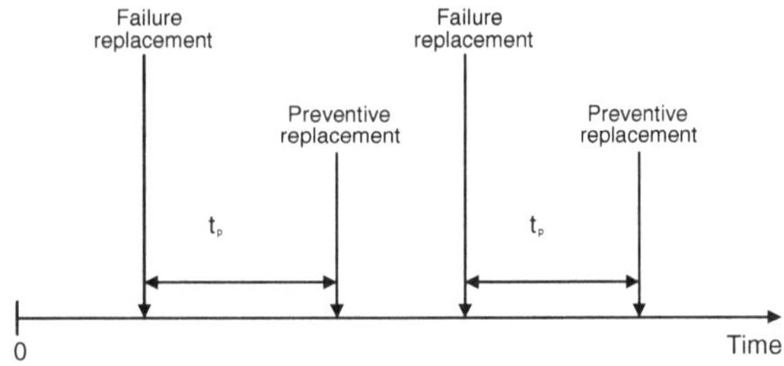

Fig. 3.5 Policy I cycles

after any type of maintenance (replacement) is performed. This policy is suited for simple equipment or a single unit in which repair at the time of failure (or replacement) could nearly correspond to general overhaul. An example of such equipment is a vacuum tube. This policy is illustrated in Fig. 3.5.

In this situation, as shown in Fig. 3.5, there are two operations cycles. In one cycle, the equipment operates till the time for preventive maintenance (replacement) $t_{p,}$ and in the second cycle, the equipment fails prior to the planned maintenance.

The objective of the model in this section is to determine the optimal t_p, meaning that the t_p at which preventive replacement is performed after the equipment has operated continuously for t_p hours without failure. The model determines the t_p that minimizes the total expected cost of preventive and breakdown maintenance per unit time shown in Eq. 3.2.

$$\mathrm{UC}(t_p) = \frac{\text{Total expected cost per cycle}}{\text{Expected cycle length}} \tag{3.2}$$

The total expected cost per cycle consists of the cost of preventive maintenance in addition to the cost of breakdown (failure) maintenance, which is

$$C(t_p) = C_p \times R(t_p) + C_f\left[1 - R(t_p)\right] \tag{3.3}$$

Where $R(t_p)$ is the probability the equipment survives till age t_p, which is represented by the shaded area in Fig. 3.6.

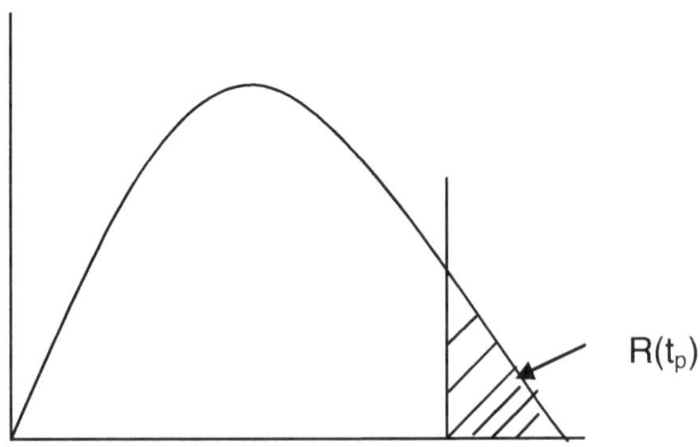

Fig. 3.6 Area under the probability distribution representing $R(t_p)$

The expected cycle length consists of the expected length of preventive cycle plus the expected length of a failure cycle.

$$\text{Expected Cycle Length} = t_{\mathrm{p}}R(t_{\mathrm{p}}) + M(t_{\mathrm{p}})(1 - R(t_{\mathrm{p}})) \tag{3.4}$$

where

$$M(t_{\mathrm{p}}) = \int_{-\infty}^{\infty} tf(t)\mathrm{d}t/[1 - R(t_{\mathrm{p}})] \tag{3.5}$$

$M(t_{\mathrm{p}})$ is the mean of the truncated distribution at t_{p} (see Fig. 3.6)

$$\mathrm{UC}(t_{\mathrm{p}}) = \frac{C_{\mathrm{p}}R(t_{\mathrm{p}}) + C_{\mathrm{f}}[1 - R(t_{\mathrm{p}})]}{t_{\mathrm{p}}R(t_{\mathrm{p}}) + M(t_{\mathrm{p}})[1 - R(t_{\mathrm{p}})]} \tag{3.6}$$

$\mathrm{UC}(t_{\mathrm{p}})$ is a function of one variable which is $t_{\mathrm{p}.}$ Line search methods such as the golden section method or Newton method can be utilized for solving this type of problem. The next section provides the details of the golden section method.

3.6.2 The Golden Section Method

The algorithm in this section is a direct search algorithm. It depends on functional evaluation and uses no information about derivatives. Other derivative-based methods such as the Newton method given in Appendix A can be utilized to solve problems that seek the minimum or maximum of a single variable function. The algorithm requires the problem in the following form

$$\begin{aligned} &\text{Minimum } g(t) \\ &\text{Subject to } a \le t \le b \end{aligned} \tag{3.7}$$

The steps of the algorithm are as follows:

1. Choose an allowable final tolerance level, δ, and assume that the initial interval where the minimum lies is $[a_1, b_1] = [a, b]$ and let $\lambda_1 = a_1 + (1 - \alpha)(b_1 - a_1)$, $\mu_1 = a_1 + \alpha(b_1 - a_1)$, and $\alpha = 0.618$. Evaluate $g(\lambda_1)$ and $g(\mu_1)$, let $k = 1$, and go to step 2.
2. If $b_k - a_k < \delta$, stop as the optimal solution is $t^* = \frac{a_k + b_k}{2}$. Otherwise, if $g(\lambda_k) > g(\mu_k)$, go to step 3, and if $g(\lambda_k) \le g(\mu_k)$, go to step 4.
3. Let $a_{k+1} = \lambda_k$ and $b_{k+1} = b_k$; furthermore, let $\lambda_{k+1} = \mu_k$ and $\mu_{k+1} = a_{k+1} + \alpha(b_{k+1} - a_{k+1})$, evaluate $g(\mu_{k+1})$, and go to step 5.

4. Let $a_{k+1} = a_k$ and $b_{k+1} = \mu_k$; furthermore, let $\mu_{k+1} = \lambda_k$ and $\lambda_{k+1} = a_{k+1} + (1 - \alpha)(b_{k+1} - a_{k+1})$, evaluate $g(\lambda_{k+1})$, and go to step 5.
5. Replace k by $k + 1$ and go to step 1.

For more on the properties and the convergence of the above algorithm, see Bazarra et al. [4].

Example 1 An equipment has a time to failure density function $f(t)$ that follows a uniform distribution between [0, 10] weeks. The cost of preventive replacement is \$5, and the cost of failure replacement is \$50. Determine t_p, the optimal time of preventive replacement.

The use of these types of policies is appropriate when the failure rate function is an increasing function of time. In other words, as the equipment ages, it is more likely to fail.

The uniform probability density function for time to fail is given as

$$\begin{aligned} f(t) &= \tfrac{1}{10} \quad 0 \leq t \leq 10 \\ &= 0 \quad \text{otherwise} \end{aligned} \tag{3.8}$$

The distribution function for time t failure is

$$\begin{aligned} F(t) &= \int_0^t 1/10 \mathrm{d}t = \tfrac{1}{10}t \quad 0 \leq t \leq 10 \\ &= 1 \qquad\qquad\qquad \text{if } t \geq 10 \end{aligned} \tag{3.9}$$

The reliability of the equipment at time t is

$$R(t) = 1 - F(t) = 1 - 1/10t \tag{3.10}$$

The failure function $r(t) = \frac{f(t)}{1-F(t)} = \frac{f(t)}{R(t)} = \frac{\frac{1}{10}}{1-1/10t} = \frac{1}{10-t}$.

The term $r(t)$ is an increasing function in the specified range. In order to apply the model in this section, we need to compute $M(t_p)$, from Eq. (3.5),

$$\begin{aligned} M(t_p) &= \int_0^{t_p} tf(t)\mathrm{d}t/[1 - R(t_p)] \\ &= \int_0^t t\frac{1}{10}\mathrm{d}t/[1 - R(t_p)] = \frac{t_p}{2} \end{aligned} \tag{3.11}$$

The expected cost per unit time is

$$\begin{aligned}\mathrm{UC}(t_\mathrm{p}) &= \frac{C_\mathrm{p}R(t_\mathrm{p}) + C_\mathrm{f}\left[1 - R(t_\mathrm{p})\right]}{t_\mathrm{p}R(t_\mathrm{p}) + M(t_\mathrm{p})\left[1 - R(t_\mathrm{p})\right]} \\ &= \frac{5\left(1 - 1/10t_\mathrm{p}\right) + 50\frac{1}{10}t_\mathrm{p}}{t_\mathrm{p}\left[1 - 1/10t_\mathrm{p}\right] + \frac{t}{2}\frac{1}{10}t_\mathrm{p}} \\ &= \frac{5 + 4.5t_\mathrm{p}}{t_\mathrm{p} - \frac{1}{20}t^2p}\end{aligned} \tag{3.12}$$

The optimal t_p^* is the value of t_p that minimizes $\mathrm{UC}(t_\mathrm{p})$ given in Eq. (3.12). A simple way of doing that is to restrict our period to whole weeks and evaluate the function at the points 1, 2, …, 10 and select the value that provides the smallest value of $\mathrm{UC}(t_\mathrm{p})$. This approach is simple; however, it does not guarantee the minimum and sometimes tedious, if the number of points where the function to be evaluated is large. If a more precise solution is needed, an optimization method such as the golden section method must be employed.

Using the simple way proposed above, we evaluate the function $\mathrm{UC}(t_\mathrm{p})$ at the points 1, 2, …, 10. The result of the evaluations calculated to two decimals is shown in Table 3.2.

The minimum value is at $t_\mathrm{p}^* = 4$ weeks, i.e., the time to conduct a preventive replacement is every 4 weeks. Next, the golden section search is applied to obtain the optimal t_p^*.

Iteration 1

$$[a_1, b_1] = [0, 10], \quad \alpha = 0.618, \quad 1-\alpha = 0.382$$
$$\lambda_1 = 0 + 0.382 * 10 = 3.82 \quad \mu_1 = 0 + 0.618 * 10 = 6.18$$
$$\mathrm{UC}(\lambda_1) = \mathrm{UC}(3.82) = \frac{22.19}{3.09038}7.18, \quad \mathrm{UC}(\mu_1) = \mathrm{UC}(6.18) = \frac{32.81}{4.27038} = 7.68$$

Since $\mathrm{UC}(\lambda_1) \le \mathrm{UC}(\mu_1)$, the next interval is $[a_2, b_2] = [0, 6.18]$.

Iteration 2

$$[a_2, b_2] = [0, 6.18]$$
$$\lambda_2 = 0 + 0.382 \times 6.18 = 2.36, \quad \mu_2 = 3.82$$
$$\mathrm{UC}(\lambda_2) = \mathrm{UC}(2.36) = \frac{15.62}{2.08152} = 7.50, \quad \mathrm{UC}(\mu_2) = \mathrm{UC}(3.82) = 7.18$$

Table 3.2 Evaluation of the expected cost per unit time

t_p	1	2	3	4	5	6	7	8	9	10
$C_*^1(t_\mathrm{p})$	10	7.78	7.25	7.18	7.33	7.62	7.91	8.54	9.91	∞

Since UC(λ_2) > UC(μ_2), the next interval where the optimal solution lies is [a_3, b_3] = [2.36, 6.18].

Iteration 3

$$
\begin{aligned}
&[a_3, b_3] = [2.36, 6.18] \\
&\lambda_3 = 3.82 \quad \mu_3 = 2.36 + 0.618 \times 3.82 = 4.72 \\
&\mathrm{UC}(\lambda_3) = \mathrm{UC}(3.82) = 7.18, \quad \mathrm{UC}(\mu_3) = \mathrm{UC}(4.72) = \frac{26.24}{3.60608} = 7.28
\end{aligned}
$$

Since $\mathrm{UC}(\lambda_2) \leq \mathrm{UC}(\mu_3)$,, the next interval is [2.36, 4.72].

Iteration 4

$$
\begin{aligned}
&[a_4, b_4] = [2.36, 4.72] \\
&\lambda_4 = 3.261, \quad \mu_4 = 3.82 \\
&\mathrm{UC}(\lambda_4) = \mathrm{UC}(3.261) = \frac{19.67}{2.72862} = 7.21, \quad \mathrm{UC}(\mu_4) = \mathrm{UC}(3.82) = 7.18
\end{aligned}
$$

The next interval is [a_5, b_5] = [3.26, 4.72].

Iteration 5

$$
\begin{aligned}
&[a_5, b_5] = [3.26, 4.72] \\
&\lambda_5 = 3.82, \quad \mu_5 = 4.16 \\
&\mathrm{UC}(\lambda_5) = \mathrm{UC}(3.82) = 7.18, \quad \mathrm{UC}(\mu_5) = \mathrm{UC}'(4.16) = \frac{23.72}{3.29} = 7.21
\end{aligned}
$$

The next interval where the solution lies is [3.26, 4.16]. We can proceed to find the optimal solution. However if we decide to stop after 5 iterations, the optimal solution is approximated by:

$$
t_p^* = \frac{4.16 + 3.26}{2} = 3.71, \quad C'(3.71) = 7.179
$$

which is slightly lower than minimum UC(4), obtained by systematic search. If we approximate 3.71 to the nearest integer, we obtain $t_{\mathrm{p}}^* = 4$. Using systematic search or golden section search, we obtained almost the same solution; however, the golden section method performs an intelligent search, is more accurate than a systematic search, and usually requires less computation than a systematic search. In addition, it is available in most optimization software libraries.

3.6.3 *Optimal Constant Interval Preventive Replacement Policy (Type II Policy)*

Policy II (Block replacement) is defined as follows: perform preventive maintenance on the system after it has been operating a total of t_p hours regardless of the number of intervening failures. In case failure happens prior to $t_{p,}$ minimal repair is performed. Minimal repair does not change the failure rate of the system, and preventive maintenance renews the system and becomes as good as new. This policy is suited for complex system such as engines and turbines (Fig. 3.7).

The objective of the model in this section is to determine the optimal t_p^* that minimizes the expected repair and preventive maintenance. As mentioned earlier in this model, it is assumed that preventive maintenance brings the system to as good as new and minimal repairs do not affect the system failure rate:

$$\mathrm{UC}(t_p) = \frac{\text{Total expected cost due to preventive maintenance and minimal repair}}{\text{Length of interval}}$$

The total expected cost $C(t_p)$ consists of the cost of preventive maintenance C_p in addition to the cost of repairs. The cost of repairs is the cost of a single repair multiplied by the expected number of repairs.

$$\mathrm{UC}(t_p) = \frac{C_p + C_f H(t_p)}{t_p} \tag{3.13}$$

It has been shown by Barlow and Hunter that the expected number of failures in the interval (0, t_p) is the integral of the failure rate function, i.e.,

$$E[N(t_p) = H(t_p) = \int_0^{t_p} r(t)\mathrm{d}t \tag{3.14}$$

$UC(t_p)$ is a function of a single variable $t_{p,}$ and a direct search method such as the golden section method can be used to find t_p^*

Fig. 3.7 Policy II cycle

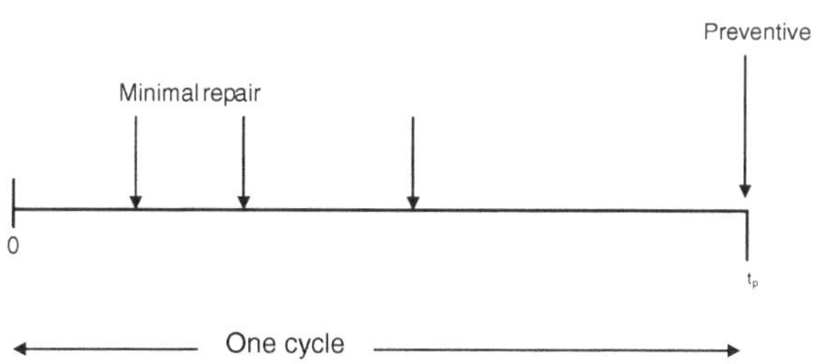

Example 2 Compute the optimal type II policy for the problem given in example 1 in Sect. 3.7.2.

Type II policy determines t_p^*, which is the number of total operating hours after which preventive maintenance is performed. The term t_p^* minimizes the following function:

$$UC(t_p) = \frac{C_p + C_f H(t_p)}{t_p} \tag{3.15}$$

We need to determine $H(t_p)$ which is the expected number of failures in the interval $(0, t_p)$. This can be obtained by using Eq. (3.14).

$$\begin{aligned}
H(t_p) &= \int_0^{t_p} r(t)\mathrm{d}t \\
r(t) &= \frac{f(t)}{1 - F(t)}, \quad \text{which is the failure rate function} \\
f(t) &= \frac{1}{10} \quad 0 \le t \le 10 \\
r(t) &= \frac{\frac{1}{10}}{1 - 1/10t} = \frac{1}{10 - t} \\
H(t) &= \int_0^{t_p} \frac{1}{10 - t}\mathrm{d}t = -\ln(10 - t)_0^{t_p}\Big]_0^{t_p} = \ln\frac{10}{10 - t_p} \\
UC(t_p) &= \frac{5 + 50\ln\frac{10}{10 - t_p}}{t_p} = \frac{5}{t_p} + \frac{50}{t_p}\ln\frac{10}{10 - t_p}
\end{aligned} \tag{3.16}$$

Applying the simple systematic search procedure used in the previous example for the function $UC(t_p)$ above, for the integer values in the range (0, 10), *the values in* Table 3.3 are obtained.

The value of t_p that provides the least cost is $t_p^* = 3$. Therefore, preventive maintenance should be performed every 3 weeks. It is better to apply the golden section methods to find the optimal t_p^* without any restrictions on t_p as done in the previous example.

Table 3.3 Expected cost evaluation for example 2

t_p	1	2	3	4	5	6	7	8	9	10
$UC(t_p)$	10.27	8.07	7.61	7.63	7.93	8.47	9.30	9.57	13.35	∞

3.6.4 Extension of Policies I and II

Nguyen and Murthy [14] generalized the two basic preventive maintenance policies proposed by Barlow and Hunter. Their type I policy is suited for a single system such as an engine or a television, and their type II policy is suited for a multi-item system such as bulbs in a factory. The policies are described as follows:

Policy I is defined as follows: replace the system after $(k - 1)$ repairs. For a system subjected to $(i - 1)$ repairs, it is repaired (or replaced if $i = k$) at time of failure or at age T_i (T_i = hours from last repair or replacement) whichever occurs first.

Policy II is defined as follows: replace the system after $(k - 1)$ repairs. For a system subjected to $(i - 1)$ repairs, it is always repaired (or replaced if $i = k$) at age T_i. In case of failure, a minimal repair is carried out.

In case $k = 1$, the above two policies reduce to the one proposed by Barlow and Hunter and presented in the previous sections. The above two policies are characterized by k and T_i, $i = 1, \ldots, k$ where T_i is the maintenance age.

3.6.4.1 Optimal Type I General Policy

In the following model, the maintenance costs consist of replacement cost C_{R}, preventive maintenance C_{p}, and breakdown (failure) maintenance cost C_{f}.

For a system subjected to $(i - 1)$ preventive maintenance, if we adopt the policy of performing repair at failure or at age T_i, whichever occurs first, the expected cost of preventive maintenance is given as follows:

$$C_{\mathrm{p}}(T_i) = C_{\mathrm{p}} + C_{\mathrm{f}} F_i(T_i) \tag{3.17}$$

Therefore, the expected cost of replacement cycle is

$$C(k, T_i) = (k - 1)C_{\mathrm{p}} + C_{\mathrm{R}} + C_{\mathrm{f}} \sum_{i=1}^{k} F_i(T_i) \tag{3.18}$$

The expected cycle length is given as follows:

$$L(k, T_i) = \sum_{i=1}^{k} \int_0^{T_i} \overline{F}_i(t)\mathrm{d}t \tag{3.19}$$

Since it is known that

$$\mu_i = \int_0^{T_i} \overline{F}_i(t)\mathrm{d}t \tag{3.20}$$

where μ_i is the expected value of T_i.

The model that minimizes the expected cost per unit time is

$$C[k, T_1, \ldots, T_k] = \frac{\left[(k-1)C_p + C_R + C_f \sum_{i=1}^{k} F_i(T_i)\right]}{\sum_{i=1}^{k} \int_0^{T_i} \overline{F}_i(t)\mathrm{d}t} \tag{3.21}$$

The optimal policy is to select k and maintenance ages $\{T_i(k)\}$, $i = 1, \ldots, k$, so as to minimize $C[k, T_i]$ in Eq. (3.21).

3.6.4.2 Optimal Type II General Policy

In the case of the general type II policy, the maintenance costs consist of replacement cost, preventive maintenance cost, and minimal repairs (breakdown maintenance).

The expected total cost per unit time for the general type II policy is given as

$$C[k, T_1, \ldots, T_k] = \frac{(k-1)C_p + C_R + C_f \sum_{i=1}^{k} \int_0^{T_i} r_i(t)\mathrm{d}t}{\sum_{i=1}^{k} T_i} \tag{3.22}$$

where $r_i\ (t)$ is the failure rate at time t for a system subject to $(i - 1)$ repairs.

This cost function in (3.22) can be minimized using optimization techniques to obtain the optimal policy.

3.7 Inspection Models

The main purpose of inspections is to obtain useful information about the state of an equipment or larger technical systems. Inspectors collect information about useful indicators such as bearing wear, gauge readings, vibrations, oil debris, and the quality of the product. Information on these indicators can be used to predict equipment failure and plan further maintenance actions depending on the state of the equipment. Inspections are useful and can lead to the following:

- less extensive repairs of potential failure if it is detected before it creates follow-up damage.
- Proper planning of corrective actions so they can be performed at times when they cause the least disruption to the system operations.

CBM has been widely accepted in the past few years, since it enables maintenance decisions to be made based on the state of the equipment. This leads to more cost-effective maintenance by reducing unnecessary repairs, overhauls, and replacements. The CBM method relies on monitoring and the analysis of data obtained from an operating system. Continuous monitoring and inspection provide

CBM with the necessary information about the state of the equipment. The frequency of inspection and the level of monitoring depend on the cost of inspection and the benefit of the inspection in providing correct information about the state of the system and the ability to predict the eminence of failure from the state. Therefore, inspection decisions are important for the success of any predictive maintenance program. In this section, several models that aid in determining the frequency of inspection will be presented.

3.7.1 Optimal Inspection Schedule Minimizing Expected Cost for a Single Machine

An equipment is used as a part of a production process. The equipment is either in a good or failure state. It is possible to detect the state of the equipment through inspection of its product or other indicators. When failure is detected, the equipment is returned to a good state through maintenance and the production cycle starts again in an infinite horizon. The model in this section will determine the optimal inspection schedule that minimizes the total expected cost per unit time associated with inspection, maintenance, and non-detection of failed equipment. This is the same model in [10]. In some sense, this model is equivalent to replacing the equipment when fault is detected. The inspection policy is to conduct an inspection at times $x_1, x_2, x_3, \ldots, x_n$ until failure is detected. When failure is detected, the equipment is brought to a new condition through maintenance and the production cycle begins. The horizon we have is infinite.

The model employs the following notations:

$f(t)$	the density function of the time to failure of the equipment
C_{i}	the cost of inspection
C_{u}	the cost per unit associated with undetected equipment failure
C_{r}	the cost of repair
T_{r}	the repair time
$\mathrm{EC}(x_1, x_2, \ldots, x_n)$	the total expected cost per cycle
$\mathrm{ET}(x_1, x_2, \ldots, x_n)$	the expected cycle length
$\mathrm{UEC}(x_1, x_2, \ldots, x_n)$	the total expected cost per unit time

The objective is to determine $x_1, x_2, \ldots, x_n$ that minimizes $\mathrm{UE}(x_1, x_2, x_3, \ldots, x_n)$. If failure happened between x_{i-1} and x_i, at time t_i, the cost of the cycle would be

$$iC_i + C_{\mathrm{u}}(x_i - t_i) + C\underline{\mathrm{r}} \tag{3.23}$$

and the expected value of the cost is

$$\int_{x_{i-1}}^{x_i} [iC_i + C_u(x_u - t) + C_r)]f(t)\mathrm{d}t \tag{3.24}$$

Summing over all i, we obtain the expected cost $E_i(x_1, x_2, x_3, \ldots, x_n)$ in Eq. (3.25):

$$\begin{aligned} \mathrm{EC}(x_1, x_2, x_3, \ldots, x_n) &= \sum_{k=0}^{\infty} \int_{x_k}^{x_{k+1}} [(k+1)C_i + C_u(x_{k+1} - t) + C_r]f(t)\mathrm{d}t \\ &= C_r + \sum_{k=0}^{\infty} \int_{x_k}^{x_{k+1}} [(k+1)C_i + C_u(x_{k+1} - t)]f(t)\mathrm{d}t \end{aligned} \tag{3.25}$$

Using similar arguments, the expected cycle length is

$$\mathrm{ET}(x_1, x_2, x_3, \ldots, x_n) = \mu + T_r + \sum_{k=0}^{\infty} \int_{x_k}^{x_{k+1}} (x_{k+1} - t)f(t)\mathrm{d}t \tag{3.26}$$

The expected cost per unit time is obtained by dividing the term in Eq. (3.25) by the one in (3.26) and is given as

$$C'(x_1, x_2, x_3, \ldots, x_n) = \frac{C_r + \sum_{k=0}^{\infty} \int_{x_k}^{x_{k+1}} [(k+1)C_i + C_u(x_{k+1} - t)]f(t)\mathrm{d}t}{\mu + T_r + \sum_{k=0}^{\infty} \int_{x_{k-1}}^{x_{k+1}} (x_{k+1} - t)f(t)\mathrm{d}t} \tag{3.27}$$

Other models that maximize availability are given in detail in [10].

3.7.2 A Profit Maximization Model for a Single Machine Inspection

In this section, the frequency of inspection for a machine subject to failure is determined using a model that maximizes a profit per unit time. This model has been developed by Hariga in [9]. The model is formulated for a machine that is used in a production process. The machine has a general failure distribution. Inspections will reveal the condition of the machine and may result in reducing the severity of failure. A cost C_r is incurred for repairing a failed machine. The cost of inspection is C_i and p is the profit per unit time. The question the model answers is how often this machine should be inspected to maximize profit?

The expected profit per cycle $P(T)$ consists of a profit $P_1(t)$ from a cycle without failure multiplied by the probability that failure does not happen in the cycle plus a profit $P_2(T)$ of a cycle with a failure multiplied by the probability of failure in the cycle.

For a cycle without failure, $P_1(t)$ is given as

$$P_1(T) = pT - C_i \tag{3.28}$$

For a cycle with failure, assume failure occurred at t, $t < T$. Then,

$$P_2(T) = E[pt|t<T] - C_i - C_r = \frac{\int_0^T pt(f(t)\mathrm{d}t}{F(T)} - C_i - C_r \tag{3.29}$$

The expected profit per cycle is obtained by multiplying the terms in Eqs. (3.28) and (3.29) by their respective probabilities and adding them as given in Eq. (3.30).

$$P(T) = (pT - C_i)R(T) + \int_0^T ptf(t)\mathrm{d}t - (C_i + C_r)F(T) \tag{3.30}$$

Substituting $F(T) = 1 - R(T)$, we set

$$P(T) = p\int_0^T R(t)\mathrm{d}t + C_r R(T) - C_i - C_r \tag{3.31}$$

The expected profit per unit time $UP(t) = P(T)/T$ is given as

$$UP(T) = \left(p\int_0^T R(t)\mathrm{d}t + C_r R(T) - C_i - C_r\right)/T \tag{3.32}$$

$UP(T)$ is a function of one variable that needs to be maximized.

3.7.3 Inspection Model to Minimize Expected Cost with Minimal Repair

A single machine is used in a production process. Each production cycle begins with a new machine (or one overhauled to the condition good as new). After a production period, the process may shift to an out-of-control state which is equivalent to machine failure. When the process shifts to an out-of-control state, a maintenance action is performed (minimal repair) on the machine to bring the process under control. Inspection is performed at times $T_1, T_2, \ldots, T_n$ to observe the machine (process) in order to take appropriate maintenance action. Three types of cost are considered; these are as follows: the replacement or overhaul cost C_r, the cost of inspection C_i, and the cost of repair C_f and the increased cost per unit time of running in an out-of-control state s. The objective is to determine the optimal value of n and T_i, $i = 1, 2, \ldots, n$ such that the expected total operating cost per unit time is minimized. The assumption of the model is as follows:

1. The life of the machine is a random variable with probability density function $f(t)$.
2. Repair times are negligible, and the repair brings the machine back to an in-control state. The repairs do not change the failure distribution (i.e., the repairs are minimal).

The model in this section differs from the one in Sect. 3.8.1 because it includes minimal repair, inspection, and replacement. Also, it relates the cost of failure to the process being out of control in a more direct way.

The total expected cost during a production cycle T consists of (1) replacement cost, C_r; (2) inspection cost, nC_I; and (3) the expected cost due to repair and running in the out-of-control state during the production cycle. The last cost consists of a sum of n terms (one from each inspection interval) resulting from repair and running in the out-of-control state during the ith inspection interval. Clearly, $i = 0, 1, 2, 3, \ldots, n - 1$.

$$C_{i+1} = \int_{T}^{T_{i+1}} [r + s(T_{i+1} - y)]f(t)\mathrm{d}t/\overline{F}(T_i) \tag{3.33}$$

where

$$F(T_i) = 1 - F(T_i), \quad \text{and} \quad F(T_i) = \int_0^{T_i} f(t)\mathrm{d}t \tag{3.34}$$

Then, the expected cost during a production cycle is obtained by dividing the term in (3.34) by T and is given as:

$$E(C) = C_r + nC_i + \sum_{i=0}^{n-1} C_{i+1} \tag{3.35}$$

The expected cost per unit time is given as

$$E'[C] = \left[C_r + nC_i + \sum_{i=0}^{n-1} C_{i+1}\right] \Big/ T$$

The model can be simplified by considering $T_n = T$, and the objective is to determine the number of inspections n and the time for each inspection. Also, it has been shown that if the failure rate follows the exponential distribution then the inspections are equally spaced, i.e.,

$$T_{i+1} - T_i = T_i - T_{i-1} \quad \text{for all } i$$

Detailed analysis of this model and further extensions are given in Banerjee and Chuiv [2].

3.7.4 A Model for Coordinating the Inspection of a Group of Machines

Ben Daya and Duffuaa [5] have extended the model in the previous section to coordinate the inspection of several machines. Coordination of maintenance activities takes advantage of setup costs and can result in savings. In this section, a model that develops a schedule of inspections for a group of machines will be presented. The cost in this model consists of two parts: a setup cost incurred whenever a review to determine which machines to inspect takes place regardless of the number of machines inspected and a setup cost associated with each machine to be inspected. The second part is the failure cost which consists of the cost of repair and the incurred cost for running in out-of-control state.

The time between two consecutive setups is the basic cycle. The coordination takes advantage of setup cost to minimize the total expected cost. The objective of the model is to determine inspection time, T_i, for machine i as a multiple of the basic cycle and minimize the expected cost per unit time.

The notations needed to state the model are the following:

N number of machines
A average setup cost which is incurred whenever a review for inspection is carried out regardless of which machines are inspected
a_i inspection cost of machine i
C_{ij} cost of failure of machine i during interval j
T_0 time between two consecutive setup times (called a basic cycle)
T_{ij} time between $(j-1)$th and jth inspections of machine i
k_i number of basic cycles between two consecutive inspections of machine i
$f_i(t)$ probability density function of the time to shift to an out-of-control state of machine i
$F_i(t)$ cumulative distribution function
r_i repair cost of machine i
s_i increased cost per unit time of running machine i in an out-of-control state
t_{ij} time of jth inspection of machine i
T length of the planning horizon

The assumptions of the model are as follows:

1. A cycle schedule is repeated every T time units.
2. At the end of every interval of length T_0, a review is made to see which machine should be inspected at cost A, called setup cost. This periodic review, at equal intervals of length $T_{0,}$ is appealing from a practical point of view since it simplifies administering the inspection activities.
3. Each machine is subject to a shift from an in-control state to an out-of-control state that can only be detected through inspection. The time to the shift of machine i is an exponential random variable with a probability density function of $f_i(t) = \frac{1}{\lambda_i} e^{-\frac{t}{\lambda_i}}$. This assumption allows us to use a constant inspection interval for each machine, i.e., $t_{i,j+1} - t_{i,j} = t_{i,j-1} = T_i$ for all j. This assumption can easily be relaxed, and the model can be generalized to encompass general failure functions.

The total expected cost consists of three types of costs that can be computed as follows:

1. Setup cost is incurred at every basic cycle and is computed as

$$T/T_0 * A \tag{3.36}$$

2. Cost of inspection for each machine is the cost of inspection for machine i, summed over all machines.

$$\sum_{i=1}^{N} \frac{T}{T_i} * a_i \tag{3.37}$$

3. Failure cost consists of the cost of repair plus the increased cost when the machine is operating in out-of-control state. The failure cost when machine is went out of control at time t in the interval $[t_{i,j}, t_{i,j+1}]$ is given as

$$C_{i,j+1} = \int_{t_{i,j}}^{t_{i,j+1}} \left[ri + s\left(t_{i,i+1} - y\right)\right] f_i(t)\mathrm{d}t / \overline{F}\left(t_{i,j}\right) \tag{3.38}$$

The total cost is obtained by summing $C_{i,j+1}$ in eq. (3.38) over all intervals for each machine and then summing over all machines. It is given as

$$\sum_{i=1}^{N} \sum_{j=0}^{n_{i-1}} C_{i,j+1} \tag{3.39}$$

where $n_i = T/T_i$

Therefore, the total expected cost per cycle T is

$$\mathrm{TEC}(T_0, T_1, T_2, \ldots, T_n) = \frac{T}{T_0} * A + \sum_{i=1}^{N} \frac{T}{T_i} * a_i + \sum_{i=1}^{N} \sum_{j=0}^{n_{i-1}} C_{i,j+1} \tag{3.40}$$

The expected cost per unit time is obtained by dividing $\mathrm{TEC}(T_0, T_1, T_2, \ldots, T_n)$ in Eq. (3.40) by T

$$\mathrm{TEC}(T_0, T_1, T_2, \ldots, T_N) = \frac{A}{T_0} + \sum_{i=1}^{N} \frac{a_i}{T_i} + \sum_{i=1}^{n-1} \frac{C_{i,j+1}}{T_i} \tag{3.41}$$

and $T_i = kT_0$, where k is an integer.

3.8 Imperfect Preventive Maintenance

The models presented so far in this chapter and those applied in most studies on repair, preventive maintenance, and overhaul are based on two assumptions. At one extreme is the assumption of "bad as old," which means that the failure rate of the system is not reduced by minimal repair. On the other extreme is the assumption of "good as new" that means that system is restored to a new condition after preventive maintenance. In most real-world situations, all types of maintenance may enhance the state of the equipment or system at a level between these two extremes and occasionally, it might be worse than before preventive maintenance due to faulty procedures.

Recently, the concept of imperfect maintenance was introduced by Malik and Nakagawa [11–13] in which it is assumed that maintenance will improve the equipment condition somewhat, but will not restore it to the new state unless the equipment is virtually completely replaced. The level of improvement depends on the amount and quality of resources used to perform the maintenance. Three approaches have been suggested to model the effect of preventive maintenance on the state of the equipment. The approaches are as follows:

1. An equipment after PM has the same failure rate as before PM or is good as new with certain probabilities.
2. The equipment age reduces by x unit of time after each PM.
3. The age and the failure rate of an equipment are reduced to the original ones at the beginning of all PM in proportion to the PM cost.

Models for the above approaches have been developed and optimum imperfect preventive maintenance policies that minimize expected cost per unit time are reported in [13].

3.8.1 Models Assumptions and Notation

The models describe a one-unit system which should operate for an infinite horizon under the following assumptions:

1. The unit starts to operate at time 0. It has a failure distribution $F(t)$ and a density function $f(t)$.
2. The failure rate $r(t) = \frac{f(t)}{1-F(t)}$ is montonically increasing.
3. The unit is maintained preventively at times kT ($k = 1, 2, \ldots, T > 0$).
4. The unit undergoes only minimal repairs at failures between preventive maintenance. Minimal repair does not change the failure rate. (This is a type II PM policy.)
5. The repair and PM times are negligible.
6. The cost of PM is C_p and the repair cost is C_f.
7. EC[] denotes the expected cost per cycle, E[CD] denotes expected cycle duration, and UE[] denotes the expected cost per unit time.

3.8.2 Approach One Model

In this approach for modeling imperfect maintenance, the equipment after PM has the same failure rate as existed before PM with probability P ($0 \leq P < 1$) and is as good as new with probability $q = 1 - P$.

The expected total cost from time zero to the time the equipment is as good as new by perfect PM is derived as follows. If the equipment survived up to the time jT, then the cost consists of the cost of j preventive maintenance and the expected number of repairs up to time jT. It is given as follows:

$$JC_p + Cf \int_0^{jT} r(t)\mathrm{d}t \tag{3.42}$$

The expected cost will be obtained by multiplying the term in Eq. (3.42) by the probability of survival up to time jT which is qp^{j-1}. Therefore, the expected cost is given by

$$E[C(P,T)] = \sum_{j=1}^{\infty} qP^{j-1}\left(jC_p + C_f \int_0^{jT} r(t)\mathrm{d}t\right) \tag{3.43}$$

The expected cycle duration up to perfect PM using the same probabilistic argument is given by

$$\mathrm{E[CD]} = \sum_{j=1}^{\infty} qP^{j-1}(jT) \tag{3.44}$$

Dividing the expected cost in Eq. (3.43) by the expected duration in Eq. (3.44) results in the expected cost per unit time

$$\begin{aligned} \mathrm{UE}[C(P,T)] &= E(C(P,T)]/E(CD) \\ &= \frac{\sum_{j=1}^{\infty} qP^{j-1}\left(jC_\mathrm{p} + C_\mathrm{f}\int_0^{jT} r(t)\mathrm{d}t\right)}{\sum_{j=1}^{\infty} qP^{j-1}(jT)} \end{aligned} \tag{3.45}$$

The above term can be simplified as follows:

$$\begin{aligned} \sum_{j=1}^{\infty} qP^{j-1}(jC_\mathrm{p}) &= qC_\mathrm{p}\sum_{j=1}^{\infty} jP^{j-1} = qC_\mathrm{p}\frac{\mathrm{d}}{\mathrm{d}p}\left(\sum_{j=1}^{\infty} P^j\right) \\ &= qC_\mathrm{p}\frac{\mathrm{d}}{\mathrm{d}p}\left[\frac{P}{1-P}\right] = \frac{qC_\mathrm{p}}{(1-P)^2} = \frac{qC_\mathrm{p}}{q^2} = \frac{C_\mathrm{p}}{q} \end{aligned}$$

Also,

$$\sum_{j=1}^{\infty} qP^{j-1}(jT) = qT\sum_{j=1}^{\infty} jP^{j-1} = \frac{qT}{q^2} = \frac{T}{q}$$

Substituting the formula in Eq. (3.45) and simplifying, we get

$$\mathrm{UE}[C(P,T)] = \left[C_\mathrm{p} + C_\mathrm{f}q^2\sum_{j=1}^{\infty} P^{j-1}\int_0^{jT} r(f)\mathrm{d}t\right] \Big/ T \tag{3.46}$$

Differentiating UE[$C(P, T)$] and setting the derivative to zero, we get

$$\sum_{j=1}^{\infty} P^{j-1}\int_0^{jT} t\mathrm{d}r(t) = C_\mathrm{p}/C_\mathrm{f}q^2 \tag{3.47}$$

It is clear that the left-hand side of the previous equation is increasing from the assumption about $r(t)$. Thus, $\int_0^{\infty} t\mathrm{d}r(t) > C_\mathrm{p}/C_\mathrm{f}q^2$ and therefore, a finite and unique T^* exists that satisfies Eq. (3.47) and the resulting expected cost is

$$\mathrm{UE}[C(T^*,P)] = C_\mathrm{p}q^2\sum_{j=1}^{\infty} P^{j-1}r(jT) \tag{3.48}$$

3.8.3 Approach Two Model

In this approach, the equipment becomes x units of time younger after each PM ($0 \le x \le T$). If $x = T$, then the equipment as good as new (perfect PM). If $x = 0$, then the equipment as bad as old. Suppose the equipment is replaced after N intervals, i.e., at age NT. N is a positive integer and a cost C_r of replacement is incurred at each replacement time. Then, the expected cost per unit time is given as follows:

$$\mathrm{UE}[C(T,N,x)] = \frac{\left[C_r + (N-1)C_p + C_f \sum_{j=1}^{N-1} \int_{j(T-x)}^{T+j(T-x)} r(t)\mathrm{d}t\right]}{NT} \tag{3.49}$$

The expected cost per unit time is decreasing in x. Therefore,

$$\mathrm{UE}[C(T,N,0)] \ge \mathrm{UE}[C(T,N,x)] \ge \mathrm{UE}[C(T,N,T)] \tag{3.50}$$

If we fix N and let T be a variable between $(0, \infty)$, a necessary condition that a finite T^* minimizes UE[$C(T, N, x)$] is that it satisfies

$$\sum_{j=0}^{N-1} \int_{j(T-x)}^{T+j(T-x)} t\mathrm{d}r(t) = \left[(N-1)C_p + C_r\right]/C_f \tag{3.51}$$

Equation (3.51) is obtained by differentiating equation (3.49) with respect to T and simplifying. Next, we fix T, and then, a necessary condition for N^* to minimize *UE* [$C(T, N, x)$] is that N^* satisfies

$$\begin{aligned} &\mathrm{UE}[C(T,\ N*\ +1,x)] \ge \mathrm{UE}[C(T,\ N*,x] \quad \text{and} \\ &\mathrm{UE}[C(T,\ N*,x)] < \mathrm{UE}[C(T,\ N*\ -1,x)] \end{aligned} \tag{3.52}$$

The above condition has been used by Nakagawa to develop optimal PM and replacement policies. The details of the development are beyond the scope of this book. The advanced reader is referred to [11–13, 15].

3.8.4 Approach Three Model

Approach two for modeling imperfect maintenance assumed that the reduction in age is independent of the PM cost. In the third approach, it is assumed that the age and the failure rate of an equipment is reduced in an amount proportional to the PM cost C_p.

In case reduction in age is assumed to be proportional to PM cost and if the equipment is in the steady state, the following equation relates the new age and the old age as follows:

$$y = \left[1 - C_p/C_0\right](y + T) \tag{3.53}$$

where $y + T$ is the age just before the PM and y is the age just after PM. Therefore,

$$y = \left[C_0/C_p - 1\right]T \tag{3.54}$$

Thus, the expected cost per unit time $UE[C(T)]$ is given by

$$UE[C(T)] = \left[C_p + C_f \int_0^T r(t + y)\mathrm{d}t\right] \Big/ T = \left[C_p + C_f \int_{[C_0/C_p - 1]T}^{C_0/C_p T} r(t)\mathrm{d}t\right] \Big/ T \tag{3.55}$$

Differentiating the above equation, we get the necessary condition for the optimal T.

$$\int_{[C_0/C_p - 1]T}^{C_0/C_p T} r(t)\mathrm{d}t = C_p/C_f \tag{3.56}$$

Given $r(t)$, C_0, C_p, and C_f, we can use the above equation to find the optimal T.

In case the failure rate of an equipment reduces to $\left[1 - C_p/C_0\right]r(y + T)$ by each PM when it was $r(y + T)$ before PM, in steady-state situation, we get

$$\left[1 - C_p/C_0\right]r(y + T) = r(y) \tag{3.57}$$

The expected cost per unit time is

$$UE[C(T, y)] = \left[C_p + C_f \int_0^T r(t + y)\mathrm{d}t\right] \Big/ T \tag{3.58}$$

If the age y is obtained from Eq. (3.53) and substituted in (3.57), then a cost equation in T *alone* will be obtained. The resulting equation can be minimized by the optimization methods presented in this chapter.

3.9 Delay Time Modeling

The delay time concept in its simplest form defines a two-stage failure process. In the first stage, a defect becomes detectable, and in the second stage, this detectable defect gives rise eventually to failure of the equipment. The period h between the time when the defect is first detectable and the time of failure is called delay time. This concept is understood by engineers, and it provides the rationale for inspection and preventive maintenance. The incorporation of inspection checks may increase system availability by identifying potential causes of failure and eliminating them. It has been possible to obtain subjective estimate of the probability density function $f(h)$ of the delay time h. The knowledge about $f(h)$ enables the construction of models of the relationship between the inspection period T and other variables such as the expected downtime or the expected operating cost per unit time.

The basic inspection model developed using the concept of delay time modeling is in [6]. In this model, it is assumed that defects arise and are fixed either as breakdown repairs or as inspection repairs (p.m. repairs), i.e., identified by inspection and repaired. As the inspection period T increases, the probability of a defect giving rise to a breakdown $P(T)$ increases. In this model, $P(T)$ has been estimated under certain assumptions and a model that estimates the expected cost per unit time was developed. Following are the assumptions of the simplest inspection model developed using a delay time concept.

1. An inspection is performed every T units of time and at costs C_{I} and requires d time units, $d \ll T$.
2. Inspections are perfect in a sense that any defects present within the plant are identified.
3. Defects that are identified at inspection will be repaired within the inspection period.
4. The initial instant at which a defect may be assumed to first arise within the plant (known as the time of origin of the fault) is uniformly distributed over time since the last inspection and independent of h. Faults arise at the rate of k per unit time.
5. The probability density function of delay time $f(h)$ is known.

Assumption (3) requires that all repairs performed subsequent to an inspection can be completed within the fixed period d, independent of the number of repairs. This is a reasonable assumption if there are sufficient maintenance staff to perform repairs simultaneously. Assumption (4) provides an estimate of the expected number of defects arising in the period T, namely kT. This ignores the downtime due to breakdowns, during which no faults would arise since the machinery is idle. However, if this downtime is small compared with T, then the error will also be small.

Suppose that a fault arising within the period $(0, T)$ has a delay time in the interval $(h, h + \mathrm{d}h)$, the probability of this event is calculated as $f(h)\,\mathrm{d}h$. This fault will be repaired as a breakdown repair if the fault arises in period $(0, T - h)$ (see Fig. 3.8), otherwise as an inspection repair. The probability of the fault arising

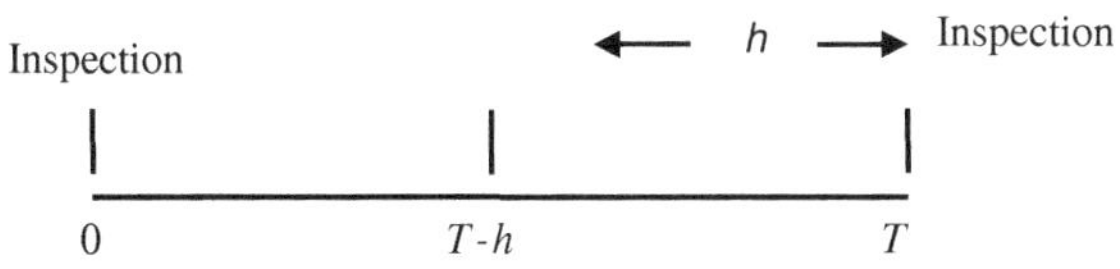

Fig. 3.8 Delay time process

before $(T - h)$, given that a fault will arise, is $(T - h/T)$ (assumption 4). The probability that a fault is repaired as a breakdown and has delay time in $(h, h + \mathrm{d}h)$ is given by

$$(T - h)/Tf(h)\mathrm{d}h \tag{3.59}$$

Summing over all possible h, the probability of a fault arising as a breakdown $P(T)$ is given by

$$P(T) = \int_{h=0}^{T} \left(\frac{T-h}{T}\right) f(h)\mathrm{d}h. \tag{3.60}$$

If the average downtime for breakdown repair is d_b, the expected downtime per unit time to be incurred under an inspection policy of period T is given by $D(T)$, where

$$D(T) = \frac{1}{(T+d)}[kTd_bP(T) + d] \tag{3.61}$$

and k is the arrival rate of defects per unit time. Equation (3.61) can be used as an objective function to determine the optimal inspection frequency T. A more appropriate objective is to minimize the expected cost which consists of the expected costs of breakdown repairs, preventive maintenance, and the cost of inspection.

The expected cost of breakdown repair $C_f(T)$ is obtained by multiplying the cost of breakdown repair C_f by the expected number of breakdown in the interval T. It is given as follows:

$$C_f(T) = C_f k\, T\, P(T) \tag{3.62}$$

The expected cost of preventive maintenance is

$$C_p(T) = C_p k\, T(1 - P(T)) \tag{3.63}$$

The cost of inspection is C_{I} . Then, the expected total cost per unit time is

$$\mathrm{UC}(T) = \left(\frac{1}{(T+d)}\right)\left[kT\{C_{\mathrm{f}}P(T) + C_{\mathrm{p}}[1 - P(T)]\} + C_{I}\right] \tag{3.64}$$

The above model has been extended to non-perfect inspection cases and to situations when assumption (3) three does not hold. The above model can be used to determine optimal inspection frequency and preventive maintenance schedules and has been applied in many case studies of maintenance of production plants. The complete detail of this topic is beyond the scope of this book. For detail, see [1, 6–8].

3.10 Summary

In this chapter, the concept of preventive maintenance is presented together with several mathematical models for inspections and preventive models. In addition, the chapter covers the classical type 1 and II polices in addition to condition-based maintenance. Also, the concept of delay time is presented and two delay models are discussed and highlighted with simple examples. Key inspection models are outlined and demonstrated with examples.

Exercises

1. Why is preventive maintenance in general is more effective than other types of maintenance strategies?
2. Visit a plant near your area and identify the need for CBM. List the diagnostic techniques the plant needs. Identify the type of monitoring and equipment needed to make the CBM operational.
3. What are the cost components involved in the equipment life cycle? How do you estimate each component?
4. Visit a nearby plant and select a critical equipment and design a planned maintenance program for it.
5. An equipment time to failure follows an exponential distribution with mean equal 500 days. The cost of preventive and failure maintenance are \$50 and \$300, respectively. Determine t_{p} the optimal time for preventive replacement using the following policies:

   (a) Type I policy.
   (b) Type II policy.

6. Solve the same problem in 7 if the failure distribution is lognormal with scale parameter $\mu = 2$ and shape parameter $\sigma^2 = 0.01$.
7. Solve the optimal t_{p} in example 2 using the golden section method.

8. Determine the optimal t_p for the problems in example 1 and 2 using Newton method.
9. If an equipment has a failure time distribution $F(t)$ and density function $f(t)$, show the following:

   (a) $\mu = \int_0^\infty \overline{F}(t)\mathrm{d}t$
   (b) $r(t) = {}^{f(t)}/_{1 - F(t)}.$
   (c) The expected number $H(t_p)$ of failures in the interval $(0, t_p)$ is $H(t_p) = \int_0^{t_p} r(t)\mathrm{d}t$

10. Develop an optimization method to find the optimal k, T_1, …, T_k for the generalized type I and type II policies in Sect. 3.6.4.
11. Consider the inspection model in Sect. 3.8.2 and show that a breakeven inspection interval exists for any failure distribution. Show that the breakeven point is unique in the cases of the exponential and Weibull distributions (results are in [9]).
12. Consider the model in Sect. 3.8.2 and derive the optimal inspection schedule for the exponential distribution with a parameter λ in terms of p, C_r, C_I, and λ. Find the schedule if $\lambda = 0.01$, $p = 500$, $C_r = 4000$, and $C_i = 50$.
13. Consider the model in Sect. 3.8.3 and show that if the failure rate distribution is exponential, then the inspections are equally spaced, i.e., $T_{i+1} - T_i = T_i - T_{i-1}$.
14. Use the model given in Sect. 2.8.4 for coordinating the inspections of five machines. The data for the machines are given below.

Machine i	$\$C_i$	λ_i	$\$r_i$	$\$s_i$
1	5	0.01	100	10
2	10	0.05	200	100
3	5	0.08	100	10
4	5	0.02	300	100
5	10	0.1	200	50

A = 25, N = 5 (Data source is reference [3])

15. Suppose the failure distribution for an equipment is Weibull with shape parameter $\alpha = 2$ and scale parameter $\beta = 50$. Then, show the following:

   (a) The failure rate function is given by, $r(t) = \beta\alpha\, t^{\alpha-1}$
   (b) Show that $r(t)$ is monotonically increasing.
   (c) Compute T^* for the models given for approaches 1 and 3 for modeling imperfect maintenance.

References

1. Baker RD, Christer AH (1994) Review of delay-time OR modeling of engineering aspects of maintenance. Euro J Oper Res 73:407–422
2. Banerjee PK, Ni C (1996) Inspection for repairable systems. IEE 28(3):1003–1010
3. Barlow RE, Hunter LC (1960) Optimum maintenance policies. Oper Res 8:90–100
4. Bazaraa M, Shetty CM, Sherali H (1993) Nonlinear programming theory and algorithms, 2nd edn. Wiley, London
5. Ben Daya M, Duffuaa SO (1997) Co-ordination of inspections for a group of machines. J Qual Maintenance Eng (JQME) 3(3):199–208
6. Christer AH, Walker WM (1984) Delay time models for industrial inspection maintenance problem. J Oper Res Soc 35(5):401–406
7. Christer AH, Wang W (1992) A model of condition monitoring of a production plant. Int J Prod Res 30(9):2199–2211
8. Christer AH, Wang W, Sharp JM, Baker RD (1997) Stochastic maintenance modeling of high-tech steel production plant. In: Christer A, Osaki S, Thomas LC (eds) stocastic modeling in innovative manufacturing, Springer, Heidelberg
9. Hariga MA A maintenance inspection model for a single machine with general failure distribution. Microelectronics and Reliability, to appear
10. Jardine AKS (1973) Maintenance, replacement and reliability. Pitman Publishing House, London
11. Malik MAK (1979) Reliable preventive maintenance scheduling. AIIE Trans 11:221–228
12. Nagakawa T (1979) Optimal policies when preventive maintenance is imperfect. IEEE Trans Reliab R-28:331–332
13. Nagakawa T (1980) A summary of imperfect preventive maintenance policies with minimal repair. Res Oper/Oper Res 14(3):249–255
14. Nguyen DG, Murthy DNP (1984) Optimal preventive policies for repairable systems. Oper Res 30:1181–1194
15. Olorunniwo FO, Izuchukwu A (1991) Scheduling imperfect preventive and overhaul maintenance. Int J Qual Reliab Manag 8(4):67–79

Chapter 4
Maintenance Work Measurement

4.1 Introduction

The effectiveness of operations management can be substantially improved if goals to evaluate and improve the performance of such operations are established and implemented. An essential and necessary prerequisite for evaluating productivity and performance is the availability of job standards. A job standard specifies an output expected of a qualified worker performing at a standard performance. A qualified worker is the one who has acquired the skill, knowledge, and other attributes necessary to carry out the work at hand to satisfactory standards of quantity, quality, and safety. A standard performance is the rate which a qualified worker will naturally achieve without overexertion, taken as an average over the working day. Normally, job standards are used to evaluate the performance of workers, facilities, and for predicting, planning, scheduling, and controlling work, costs, and operations. Job standards are necessary for planning maintenance resources, such as the manning of the maintenance department, and it would not be possible to do effective scheduling and control without reliable job standards. Maintenance job standards can be developed using several work measurement techniques.

The distinguishing characteristics of maintenance work must be appreciated prior to describing and illustrating the application of these techniques. It is unique when compared with typical production work. Much of the maintenance work occurs at random and at unexpected locations and is of varying duration. Maintenance work may require the worker to perform his job under varying environmental conditions such as temperature, illumination, and noise level. A typical production worker is not subjected to such uncertainties. This chapter presents a host of techniques for developing job standards taking into account the unique character of maintenance work. Section 4.2 covers work measurement techniques and Sect. 4.3 covers work sampling. Section 4.4 demonstrates the use of

S.O. Duffuaa and A. Raouf, *Planning and Control of Maintenance Systems*,
DOI 10.1007/978-3-319-19803-3_4

control charts for monitoring the quality of work sampling studies, and Sect. 4.5 presents predetermined motion time systems. Section 4.6 covers standard data and Sect. 4.7 presents estimating. A summary of the chapter is given in Sect. 4.8.

4.2 Work Measurement Techniques

The work measurement techniques can be broadly classified into two categories; (1) direct measurement technique, and (2) indirect measurement technique. These techniques are shown in Fig. 4.1. Time study and predetermined motion time systems (PMTS) are used to develop standard data for maintenance jobs that are needed to develop maintenance job standards. These techniques are briefly described below:

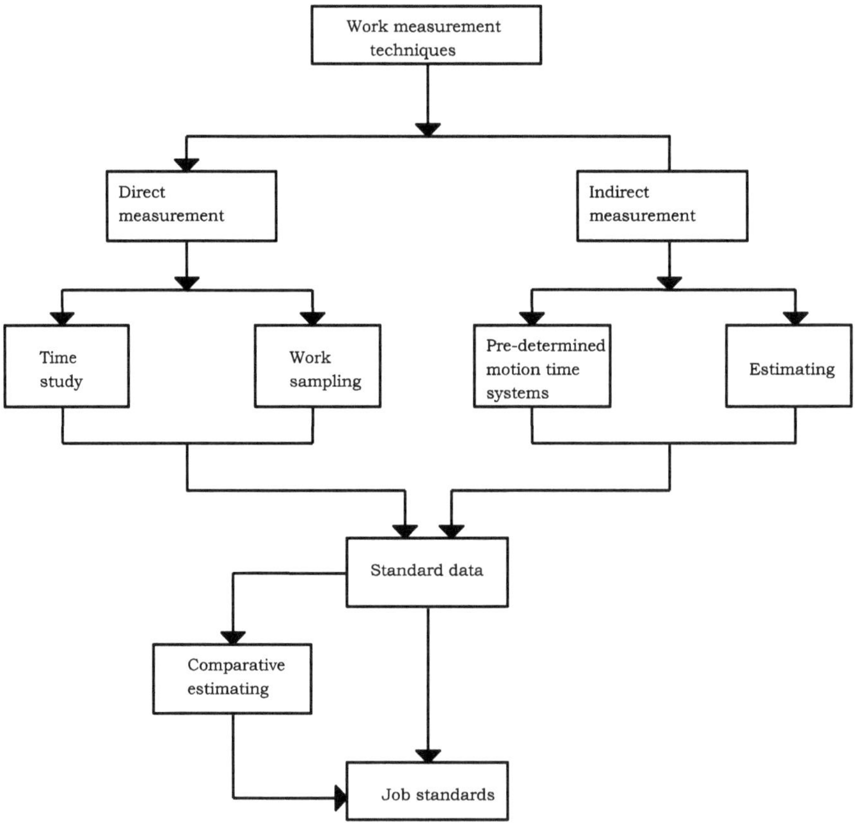

Fig. 4.1 Work measurement techniques

4.2.1 Time Study

Time study is performed by timing the worker as the job is performed, summing the times for the pertinent elements of the job, standardizing the observed times, and adding allowances for personal and other variable working conditions.

4.2.2 Time Study Procedures

Basically, the time study procedure consists of six steps as follows:

- **Select the job**. The job selected should be standardized in terms of equipment and material and the operator should be a qualified worker.
- **Break the job into elements**. Identify the work elements that constitute a work cycle. A maintenance job is divided into segments known as elements. Some of the basic considerations in breaking the job into elements include the following:
  - machine elements should be kept separate from manual elements.
  - the elements should be easy to identify and time.
  - elements associated with doing the actual maintenance work should be kept separate from obtaining material, waiting at the tool crib, walking, etc.

There are several reasons for dividing the maintenance job into elements. These include, but are not limited to the following:

- element times collected are categorized into standard data that can be combined to arrive at standards for some maintenance jobs without the need of timing these elements separately for each job.
- times for similar elements from several jobs can be compared to help keep job standards uniform.
- a maintenance worker may not perform all the elements at the same pace. The analyst must adjust the maintenance worker's time to represent the time that a qualified worker would take for each of the elements under standard performance.
- **Observing the job**. The number of times a job should be observed and timed increases with the degree of accuracy desired for the job standard and the level of significance required (the level of confidence required in estimating the job standard). The formula for finding the number of observations required n is given below.

$$n = \left(\frac{Z_{\alpha/2}\sigma}{A\mu}\right)^2 \tag{4.1}$$

where

n the total number of observations that should be taken to provide the desired accuracy
μ true mean of the time of the activity
A accuracy desired, expressed as fraction of the true value of the population parameter
$Z_{\alpha/2}$ standardized normal deviate that has $\frac{1}{2}\alpha$ as the area remaining in the tail of the standard normal distribution beyond it
σ standard deviation of the population

Since the formula in Eq. (4.1), will guarantee the required level of confidence if σ is known, and in case σ is unknown and S (sample standard deviation) is used as an estimate for σ, the following steps are necessary to find a value for n that assures the confidence level desired.

- Take an initial sample of size n_1 estimate σ and μ by S and $\overline{X}$, respectively.
- Substitute S and $\overline{X}$ (sample mean) instead of σ and μ in Eq. (4.1) and compute n.
- If $n < n_1$, stop; otherwise, let $n = n_1$ and go to step 1 above.

Example Suppose we are interested in estimating the actual mean time to perform an element. The observed element time for six observations is given below in Table 4.1.

Assume that we wish to provide 95 % confidence that the true mean time to perform this maintenance work element is estimated within 10 % accuracy. Thus,

$$\bar{X} = \frac{\sum_{i=1}^{6} x_i}{6}$$

$$S = \sqrt{\frac{\sum x_i^2 - \left(\frac{\left(\sum x_i\right)^2}{n}\right)}{n-1}} = \sqrt{\frac{0.0389 - \frac{(0.47)^2}{6}}{5}} = 0.0204$$

Table 4.1 Observed element times

Observation #	Observed element time x_i
1	0.07
2	0.08
3	0.11
4	0.09
5	0.05
6	0.07

Using Eq. (4.1), n is obtained as

$$n = \left[\frac{Z_{\alpha/2}S}{A\bar{X}}\right]^{1/2} = \left[\frac{1.96(0.0204)}{0.1(0.07833)}\right]^{1/2} = 26$$

The objective of this study is to arrive at a standard that is suitable for a qualified worker performing at standard performance; yet the worker who was timed may not have been working at a standard (100 %) pace. An adjustment known as rating or leveling is made to adjust the observed times to the time required by someone working on a "standard" pace. Rating is the process of comparing the worker's rate of performance, for each element of the maintenance job, with the analyst's concept of standard (normal) pace.

- **Compute the basic times**. A basic time for a job is the time it should take a qualified worker to perform the elements of the job while working at a standard pace. The basic time (BT) is computed using the following formula:

$$\text{BT} = \text{OT}\left(\frac{\text{Rating}}{\text{Standard Rating}}\right) \quad (4.2)$$

BT	Basic time
OT	Observed time
Standard rating	100 %

Assume a worker was observed to be working 15 % faster than what is considered standard pace. He would be given a rating of 115 %. The observed time is 4.23 min; then using Eq. (4.2), the BT is obtained as

$$\text{BT} = 4.23\left[\frac{115}{100}\right] = 4.86\,\text{min}$$

- **Determine Allowances**. An allowance is a percentage of the BT allowed for delays and fatigue. Allowances can be classified into two categories and these are constant and variable. Constant allowances are given for personal needs and fatigue. Variable allowances are related to the maintenance job characteristics. Some general guidelines for allowances are given in Table 4.2. Delays may occur on a job through no fault of the maintenance worker. He may have to wait for instructions, wait at the tool crib, or wait for a crane to come and remove a work item. An allowance for unavoidable delays has to be given.
- **Establish job standard**

$$\text{Job standard} = \text{BT}\ (1 + \text{allowances}) \quad (4.3)$$

Direct time study is generally used for developing standards for jobs requiring more than 100 h as in the case of overhauling of machines, etc. Job standards can be used

Table 4.2 General guidelines for computing allowances

Percentage to be added to the normal time for an element to make allowances for its work conditions					
A.	Constant allowances		E.	Atmospheric conditions (heat and humidity)	
	(a) Personal allowance………	5		variable………………………………	0-10
	(b) Basic fatigue allowance….	4	F.	close attention	
B.	Variable allowances			(a) Fairly fine work ………..	0
	i. Standing allowance…..	2		(b) Fine or exacting …………	2
	ii. Abnormal position allowance			(c) Very fine or exacting…………	5
	(a) Slightly awkward	0	G.	Noise level	
	(b)Awkward (bending)	2		(a) Continuous………………	0
	(c)Very awkward (lying, stretching)	7		(b) Intermittent-loud…………..	2
C.	Use of force or muscular energy (lifting, pulling, pushing)			(c) Intermittent -very loud……..	5
				(d) High-pitched-loud…………..	5
	Weight lifted (pounds)		H.	Mental strain	
				(a) Fairly complex process………..	1
	5……………………………	0		(b) Complex or wide span of attention ……….	4
	10…………………………..	1		(c) Very complex…………	8
	15…………………………..	2	I.	Monotony	
	20…………………………..	3		(a) Low…………………	0
	25…………………………..	4		(b) Medium……………..	1
	30…………………………..	5		(c) High…………………	4
	35…………………………..	7	J.	Tediousness	
	40…………………………..	9		(a) Rather tedious………………..	0
	45…………………………..	11		(b) Tedious………………………	2
	50…………………………..	13		(c) Very tedious…………………	5
	60…………………………..	17			
	70…………………………..	22			
D.	Bad light				
	(a) Slightly below recommended	0			
	(b) Well below………	2			
	(c) Quite inadequate………	5			

as a basis for developing standard data for maintenance tasks which contain maintenance job(s) for which the standards have been developed. Standard data are explained in Sect. 4.6.

4.3 Work Sampling

Work sampling is a technique of finding the percentage occurrence of a certain activity using statistical sampling. In order to obtain a complete and accurate picture of the productive time of the maintenance crafts in a certain area, it would be necessary to observe continuously all the workers in that area and to record when and why any of the workers were idle. It would be quite impossible to do this and very expensive.

Assume it is possible to note at a glance the state of each worker in a department at a given moment, and suppose for example that at one moment 70 % are observed working and 30 % are idle. If this action is repeated 50 times or more at different times of the day and each time the proportion of crafts working is always 70 %, it would be possible to say with some confidence that at any one time there were 70 % of the crafts working. As it is not generally possible to do that either, the next best method is to make tours of the maintenance department at random intervals, noting which crafts are working and which are idle, and reasons for being idle. This is the basis of work sampling, when the sample size is large and the observations are really random, there is quite a high probability that these observations will reflect the real situation plus or minus a certain margin of error.

To demonstrate the application of work sampling, assume it is desired to estimate the utilization of a maintenance craft. The craft is observed 1000 times in a six-week period at random times of the day. The work sampling record sheet below summarizes the data for the study.

The fraction of the time, p, the craft was idle can be estimated as

$$\hat{p} = \frac{X}{n} \tag{4.4}$$

where

X number of times the craft observed idle
N total number of observations

For the data in Table 4.3, $\hat{p} = \frac{250}{1000} = 0.25$

The number of times the craft observed idle, X is a binomial random variable with a probability mass function as follows:

$$b(n, p, x) = \binom{n}{x} p^x q^{n-x}, \quad x = 0, 1, 2, \ldots, \text{n} \tag{4.5}$$

$$q = 1 - p$$

The mean μ and the variance σ^2 of the random variable X are

$$\mu_X = np \quad \text{and} \quad \sigma_X^2 = npq \tag{4.6}$$

Table 4.3 A simple work sampling record sheet

Date: …	Observer: …	Study No. …
Number of observations = 1000	Total	%
Number of times craft observed working	750	75
Number of times craft observed idle	250	25

Also, $\hat{p} = \frac{X}{n}$ is a random variable with mean $\mu_{\hat{p}}$ and variance $\sigma^2_{\hat{p}}$ given as

$$\mu_{\hat{p}} = p \quad \text{and} \quad \sigma^2_{\hat{p}} = \frac{pq}{n} \tag{4.7}$$

If n is large, the distribution of $\hat{p}$ can be approximated by a normal distribution with mean p and variance pq/n, and therefore, a $(1-\alpha)$ confidence interval from p is given by

$$\hat{p} \pm Z_{\alpha/2}\sqrt{\frac{\hat{p}\hat{q}}{n}} \tag{4.8}$$

where $\hat{q} = 1 - \hat{p}$.

4.3.1 Number of Observations Required for a Work Sampling Study

The number of observations needed can be based on the standard error of p and is given by

$$N = \left[\frac{Z_{\alpha/2}}{A}\right]^2 \left[\frac{1-p}{p}\right] \tag{4.9}$$

$Z_{\alpha/2}$ and A are defined in Eq. (4.1)

Example Consider the case where the utilization of a single maintenance craft is desired. The data in Table 4.3 are collected. Estimate the interval where the idle time proportion p lies with 95 % confidence. Find how many additional observations are needed if the estimate is required to be within 10 % from P.

$$n = \left[\frac{Z_{.025}}{A}\right]^2 \frac{1-p}{p}$$

where $Z_{.025} = 1.96$ from the standard normal table.

$$A^2 = \frac{1.96^2}{1000}\frac{0.8}{0.2} = \frac{1.96^2}{1000}\frac{8}{2} = 0.01537$$
$$A = 0.124$$

The interval where p lies with 95 % confidence is $\hat{p} \pm A$, given as

$$0.175 \leq p \leq 0.225$$

The number of observations required for 95 % confidence and an accuracy level of 10 % is

$$n = \left(\frac{1.96}{0.1}\right)^2 \frac{0.8}{0.2} = 1537$$

The number of additional observations required is 1537 − 1000 = 537.

4.3.2 Planning a Work Sampling Study

The value of a work sampling study will be greatly affected by the care and effectiveness of planning efforts. The effort in the planning stage yields meaningful and statistically valid results. The following steps are suggested to be followed in planning and implementing a work sampling study.

- **Objectives of the study**

Define very clearly the objective of the study which should include the need and the specific information that is required.

- **Population Identification**

Determine the group of maintenance workers, machines, equipment, etc. that are to be work sampled. Include the characteristics of the population, i.e., its size and location. Population is to be categorized according to craft and should have visible identifying markings.

- **Activity Definition**

Select job activities to be observed that are clearly defined, mutually exclusive, and easy to recognize.

- **Observation Form Design**

Create a simple, easy to use, observation form which contains queries for all necessary information. A specimen is shown in Table 4.4.

- **Observation Routes Planning**

Plan observation routes that the observer has to follow in order to collect observation data. These should cover all possible population gathering and cover areas, a minimum of 67 % in each sampling round. There can be more than one route; however, a reasonable degree of randomness should be present. This can be achieved by a random starting point or a random selection of a route if multiple routes are planned.

- **Study Schedule**

Select a study period that will assure the reliability of work sampling findings. The period selected must reflect as much as possible the normal work load and conditions. Normally, a period of 10 working days should be available to conduct the study.

Table 4.4 Work sampling observation form

Day	Time	Activities							
Totals									
%									

- **Number of Observations**

Select an initial estimate of p for each activity in the study, the required accuracy and confidence level $(1 - \alpha)$. The required number of observations can be computed using the formula for the sample size in Eq. (4.9) for each activity. The largest value of n calculated should be used as it covers the accuracy requirement of all activities. In calculating the number of observations, n, it may be necessary to recalculate n several times to assure the required level of confidence as explained in Sect. 4.2.2.

- **Observations Time Schedule**

Schedule sampling observations employing, random numbers so as to avoid any bias. A string of random numbers can be generated using a calculator or a standard random number table.

- **Sample Population Preparation**

Inform the maintenance workforce to be work sampled of the study before commencing observations. This should include an explanation of the objectives of the study and the importance of working naturally without paying attention to the observer.

- **Observer Training**

Train observers to achieve the objectives of the study and minimize the errors and biases. This training should ensure a clear understanding of the objectives of the study. Complete familiarity with the definitions of activities and complete familiarity with the observation route(s) are necessary. This training may include trial runs of the work sampling study as planned. This will provide an estimate of p needed for the calculation of observations needed and also minimize uncertainties that the observer may have.

4.4 Control Charts

4.4.1 Introduction

Control charts are used to detect instability in a work sampling study. In this chart, p is plotted versus the individual samples gathered on a daily basis. On this chart, upper and lower control limits of p are indicated. If all the points fall within the limits, then it may be concluded that the study is stable. Instability can be attributed to errors in the methodology of the study or a change in the work environment. It indicates if a shift occurred and, therefore, incidents of instability have to be investigated and corrective actions taken.

4.4.2 Control Chart Construction

Control charts are usually started about 3 days into the beginning of the sampling study and kept updated as the study progresses.

$$\bar{p} = \frac{\text{Total \# of observations for that activity per day}}{\text{Total \# of observations for all activities that day}}$$

$$\text{Upper Control Limit (UCL)} = \overline{p} + \sqrt{\frac{\overline{p}(1-\overline{p})}{n}} \quad (4.10)$$

$$\text{Lower Control Limit (LCL)} = \overline{p} - 3\sqrt{\frac{\overline{p}(1-\overline{p})}{n}} \quad (4.11)$$

It is important to compute new limits whenever there is a change in $\bar{p}$. A typical control chart is shown in Fig. 4.2. Work sampling is a very commonly used technique for developing maintenance job standards and determining effectiveness of the maintenance workers.

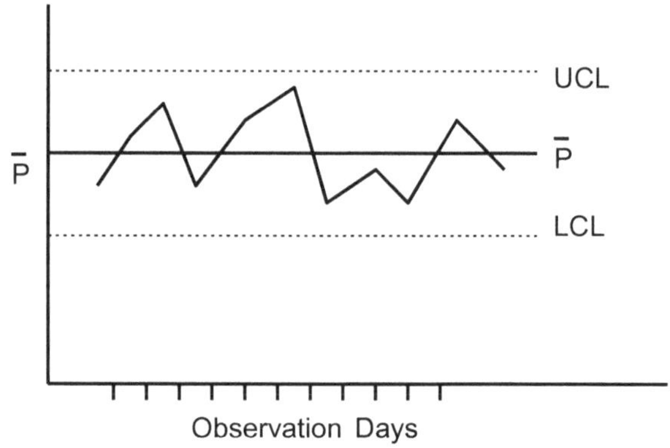

Fig. 4.2 A typical control chart

4.5 Predetermined Motion Time Systems (PMTS)

PMTS are used for setting job standards. One of the better known PMTS is the Method-Time Measurement (MTM) system). The fundamental motion times are the result of studying a large sample of diversified operations. For setting a job standard, an analyst would break the job into the basic motions required to perform it and then sum the appropriate predetermined times for all the basic motions involved. Initially, MTM was developed for production operations. These time systems in a broadened range are available now so that the system may be economical for maintenance-type operations, i.e., long cycle activities. The system is known as MTM-2.

MTM-2 System

MTM-2 is based on the categories of motions in Table 4.5.

Table 4.5 Categories of motion for MTM-2

Category	Allowed
Get	G
Put	P
Get weight	GW
Put weight	PW
Grasp	R
Apply pressure	A
Eye action	E
Foot action	F
Step	S
Bend and rise	B
Crank	C

A summary of MTM-2 data in time measurement units (TMUs) is given below. A TMU is one hundred thousandth of an hour or 10^{-5} h.

A summary of MTM-2 data (all time values are in TMUs)

Code	GA	GB	GC	PA	PB	PC
−5	3	7	14	3	10	21
−15	6	10	19	6	15	26
−30	9	14	23	11	19	30
−45	13	18	27	15	24	36
−80	17	23	32	20	30	41

GW—1 per kilogram (1/kg), PW—1/5 kg

A	R	E	C	S	F	B
14	6	7	15	18	9	61

Code	Distance
5	0–2 inch
15	≥2 < 6
30	>6 < 12
45	>12 < 8
80	≥18

The system allows for simultaneous hand motion and allows for the cases when simultaneous motion requires additional time. Effective application of these data requires proper training.

4.5.1 Procedure for Setting a Predetermined Standard

Following is the usual procedure for setting a predetermined time standard:

- Observe the job or think it through if it is yet to be established.
- Break the job into elements and record each element.
- Obtain time units for each job element from the tables.
- Add the total motion units for all elements.
- Estimate allowance for personal time, delays, and fatigue.
- Add the performance motion time and allowance units for a standard job motion unit together and compute the motion units to actual time in minutes, hours, etc. PMTS can be used for developing job standards even before the work gets started.

4.6 Standard Data

Standard data refers to standard data banks for various elements which occur repeatedly in the workplace. These elements can be put together to develop job standards.

4.6.1 Introduction

Several maintenance jobs in a company may contain the same work elements. It is not necessary to time these work elements in every job if a reasonable job standard has been determined from one or more previously studied jobs. In the maintenance department, a database can be maintained for work element duration as obtained through previous time studies or by predetermined motion time systems. If the time required for all the elements of a new job is available in standard data, they can be summed up to arrive at the basic time for the job. Personal, fatigue, and unavoidable delay allowances have to be added to the BT to arrive at a standard for the new job. Replacing of electrical motor pumps in a process industry is an example of standard data.

Standard data development is one of the desired activities of a maintenance department. Following are some of the advantages offered by it:

- eliminates repetitive work of time study analyst,
- saves time in setting job standards, and
- provides greater consistency between similar job standard.

4.6.2 Steps for Developing Standard Data

Steps for developing standard data start with gathering data to identify jobs and job standards that could be used in developing standard data. Details of jobs and pertinent element time values are put together in the form of a master list of elements. Constant value elements are separated from the variable value elements. Curve fitting and formula development using the standard techniques can be carried out.

For evaluating and improving the standard data, standard hours for the work studied are analyzed. A ratio of "time" from standard data, and "time" from "time study" for the work studied is computed. Assume time from standard data is 6.35 min and from time study is 6.44, then the index of comparison is 6.35/6.44 = 98.6. If this value is within or greater than 98 % confidence interval, then no further investigation is needed. Otherwise, standard data table values will have to be adjusted by an index of comparison where the index of comparison is given by

$$\text{I.O.C.} = \frac{\text{job standard (standard data)}}{\text{job standard (time study)}}.$$

4.7 Estimating

Estimating is the process of using past experience to predict future events. It can be used to develop job standards but in an inexpensive way.

4.7.1 Introduction

This type of work measurement is done by applying experience-based knowledge. This method is used, more or less, in every plant to a varying degree as it is not economically feasible to develop job standards for all the maintenance jobs. Some of the advantages of this method of work measurement include lower cost and being able to estimate the job standard before the job is started. However, there are some disadvantages of this method as well, and they are listed below.

- estimated job standards are often inconsistent,
- method changes may not be taken into account, i.e., the job standard may not be recomputed when necessary, and
- cannot be applied to jobs for which experience is not available.

Usually, maintenance work can be divided into the following categories:

- routine work of regular frequency,
- routine work of irregular frequency,
- one of a kind work that requires less than 100 h, and
- Jobs that require more than 100 h.

For the first two the average time taken in the past may be used as a standard. The last category may be subdivided into the following classes:

a.	0–8	hours
b.	8–16	hours
c.	16–32	hours
d.	32–50	hours
e.	50–100	hours

An average time for each such group can be used as a job standard.

4.7.2 Comparative Estimating (Slotting)

Comparative estimating is a work measurement technique used to measure the work content of low repetitive maintenance work. Setting maintenance job standards with a time study is too costly. Standard data based on a time study or PMTS offer a better solution. Maintenance job standards arrived at by estimation are inconsistent. Comparative estimating is a relatively economical alternative. It is based on the principle that an experienced craftsman can estimate a range of time during which a certain task can be completed most of the time.

4.7.3 Applying Comparative Estimating

The basis of comparative estimation is its reliance on a series of benchmark jobs. These benchmarks are jobs, representative of a range of jobs, which are similar in work requirement, types of tools used, etc., and are capable of being measured by accepted time measurement techniques.

A series of such benchmarks form "work group" or "task areas" which are similar but have a range of work content time which allows boundaries to be established within the range. Assume that a task area designated as Electric Breakdown Work covers a range of 0–60 min within the following boundaries

$$\begin{array}{ll} \text{Group D} & 0 - 20\,\text{min} \\ \text{Group E} & 20 - 40\,\text{min} \\ \text{Group F} & 40 - 60\,\text{min} \end{array}$$

The following data sheet is an example covering a particular task area. Each of the benchmarks has been measured by one of the work measurement techniques as explained earlier.

DATA-SHEET & BENCHMARKS
FOR ELECTRIC BREAKDOWN WORK
Task Area: General Electrical Repair
Group D
0-20 Minutes
Standard Time: 15 Minutes

El	Repair and adjust faulty limit switches on truck loader lifts
E2	Rectify fault on solenoid valve, traverse unit
E3	Check and rectify faults on drive motor and circuits on papering machine
E4	Repair hydraulic valve circuits on interchange unit
E5	Check and repair faulty operation of hydraulic valves on cell units

DATA-SHEET & BENCHMARKS
FOR ELECTRIC BREAKDOWN WORK
Task Area: General Electrical Repair
Group E
20-40 Minutes
Standard Time : 30 Minutes

E20	Check and repair faulty operation of lamps, switches & fuses on inspection table
E30	Repair and reset gate switches on Wadsworth lift

DATA-SHEET & BENCHMARKS
FOR ELECTRIC BREAKDOWN WORK
Task Area: General Electrical Repair
Group F
40-60 Minutes
Standard Time : 50 Minutes

E52	Repair faults on float switch
E60	Replace faulty heating element
E64	Check & repair faults on nails and relays

Group D has El, E2, E3, E4, and E5 benchmarks having standard time between 0–20 min with El closer to 0 and E5 being nearer to 20. The mean value of Group D's benchmark is 15 min. All the other groups within this task area covered by this data sheet have been similarly treated to produce a mean of 30 min for Group E and 50 min for Group F. This procedure can be followed to cover other task areas such as General Mechanical Repair. Positioning any new or unmeasured job into an appropriate group of benchmarks is known as slotting. There are two accepted methods of slotting, viz. ranking or direct comparison.

- *Ranking*: It has been said earlier that each benchmark in any particular group would be arranged in an increasing order (i.e., El on the first data sheet would be nearer to zero, while E5 would be nearer to 20 min). In the ranking system, the work specification of any unmeasured job of similar requirement would be compared with the benchmark jobs on the data sheet. When one similar but of slightly less work content than the benchmark has been found in any group, and another of slightly more work content within the same group, then the new job would be given the average time for that group.
  If the benchmark with the longer content is in the next group, then an assessment would have to be made as to whether the new job falls into, for example, Group D or Group E, or Group E or Group F.
- *Direct Comparison*: In this method, the work content of the new job is compared with a benchmark of similar work content and an assessment is made as to whether the difference between the two will move the new job out of that particular slot. The assessment is therefore whether the difference between the two jobs in work content is greater than the difference between the basic time of the benchmark job and the group or slot boundary.

Unlike production work, every maintenance activity does not require a job standard. A company can get standards by using estimation in the earlier stages and then by monitoring closely to improve the estimated job standards. The other techniques can be used when appropriate situations occur.

4.8 Summary

Numerical job standards are needed to evaluate the performance of workers and facilities, and in predicting, planning, scheduling, and controlling costs and operations. Work measurement techniques that can be used in establishing job standard in maintenance work have been discussed. The techniques include time study, work sampling, predetermined motion time study, estimating, and standard data. A step-by-step approach to be adopted in each work measurement technique is presented as well. These approaches have been illustrated with the help of examples. For more on work measurement and time standards see [1–6].

Exercises

1. A maintenance job has been studied for 20 observations. The mean actual time is 5.8 min, and the standard deviation of the time is estimated to be 2.0 min. How many additional observations should be taken for 95 % confidence that the mean actual time has been determined within 15 %?
2. An analyst has observed a maintenance job long enough to become familiar with it and has divided it into four elements. The element time for the first five cycles is shown in the following table along with the performance rating for each element.

Element no.	Cycle no. 1	Cycle no. 2	Cycle no. 3	Cycle no. 4	Cycle no. 5	Performance rating (%)
1	1.5	1.6	1.5	1.6	1.7	100
2	2.5	2.3	2.3	2.4	2.3	90
3	1.8	3.0	1.8	1.8	1.9	95
4	1.3	1.1	1.4	1.2	1.3	115

   a. Compute an estimated basic time for the job on the basis of the available data.
   b. On the basis of available data, what sample size should be taken? Estimate the time for element no. 1 within 5 % of the true mean time with 95 % confidence.

3. The analyst in problem 2 performed a work sampling study in the raw task and recorded observations as appear below in the categories of activities. What allowances should be made for unavoidable delays?

Activity category	Observed frequency
Normal work activity	112
Avoidable delays	17
Unavoidable delays	16

References

1. Aft LS (2000) Work measurement & methods improvement. Wiley, New York
2. Barnes RM (1980) Motion and time study, 7th edn. Wiley, New York
3. Benjamin NB (1988) Motion and time study, 8th edn. Howeward IL, Irwin

4. Groover M (2006) Work systems and the methods measurement and management of work. Prentice Hall, New Jersey
5. Kanawaty G (ed) (1992) Introduction to work study, 4th edn. International Labor Office, Geneva
6. Krager DW (1982) Advanced work measurement. Industrial Press, New York

Chapter 5
Maintenance Material Control

5.1 Introduction

Maintenance managers and engineers are concerned with keeping plants and machinery in adequate operating condition. To minimize emergency repairs, preventive maintenance of the equipment is carried out. The occurrence of unplanned repairs can be reduced by investigating the cause(s) of breakdown and by altering the schedule of preventive maintenance. To be able to minimize downtime, it is essential that necessary craftsmen, material, and spare parts are available.

Managers are required not only to minimize downtime but also to effectively control maintenance costs. Total maintenance costs generally comprise (1) maintenance labor cost, (2) the cost of required material and spares, and (3) the cost of production downtime when breakdowns occur. Generally, item 2 can be considered to be of the same magnitude as item 1.

One critical cost of maintenance is investment in spares and material. If the investment becomes excessive, the results are high capital costs and high maintenance costs. On the other hand, if the spares and material needed for repair and servicing the equipment are not available when needed, then the cost of downtime will increase immensely. Efforts to balance the cost of stocking maintenance material and spares against the cost of downtime are needed to achieve an effective maintenance material control system.

This chapter describes approaches for developing a maintenance material control system that will help provide the needed level of maintenance in an economical manner. A brief description of maintenance store components is provided in Sect. 5.2. The major components of the material cost are given in Sect. 5.3. Elements of maintenance store control procedure are discussed in Sect. 5.4. Effective ordering policies are provided in Sect. 5.5, followed by spares ordering policies based on failure characteristic analysis in Sect. 5.6. Section 5.7 outlines procedures for spare parts classifications based on criticality analysis followed by a brief summary of the chapter in Sect. 5.8.

S.O. Duffuaa and A. Raouf, *Planning and Control of Maintenance Systems*,
DOI 10.1007/978-3-319-19803-3_5

5.2 Maintenance Store Components

A typical maintenance store, among other categories of stores, carries spare parts, normal maintenance stock, and tools. These are defined in the following section.

5.2.1 Spare Parts

Spare parts are stocked so that equipment downtime is minimized. Spare parts may further be subdivided into the following categories:

1. Relatively expensive parts;
2. Specialized parts for use on limited number of machines;
3. Spare parts having longer than normal demand lead times;
4. Spare parts having slow turnover; and
5. Critical spare parts, the non-availability of which may cause expensive downtime or have an adverse effect on safety.

Spare parts are stocked only when the risks involved in doing without them are considered to out weigh the total cost of carrying them in stock for a predicted interval. A method of calculating this interval and the quantity is presented in the later part of the chapter.

5.2.2 Normal Maintenance Stock

This category comprises items which do not have a specialized usage, but that have a definite requirement and a short turnover. Examples of this category are commonly used bearings, pipes and pipe fittings, electric wires, switches, lumber, plywood, bolts, welding rods, etc.

Decisions regarding how much to stock and when to order for normal maintenance stocks can be handled in a more routine manner than in the case of spare parts.

5.2.3 Tools

This category usually comprises special purpose tools, which are issued on loan whenever needed.

5.3 Maintenance Material Costs

To exercise an effective cost control over maintenance operations, records must be kept concerning costs of the item, inventorying the item, and invested capital.

5.3.1 *Cost of the Item*

Cost of the item comprises the sum paid to the supplier including freight.

5.3.2 *Cost of Inventorying the Item*

Cost of inventorying the item sometimes is estimated as a percent of dollar value, expressed as a decimal fraction. Normally, it varies between 15 and 20 % of the item cost per year.

5.3.3 *Cost of the Item at Its Time of Issue*

By considering the following, costs, the cost of an item at the time of issue can be estimated:

1. Cost of space and ancillary facilities per m^2 of storage areas.
2. Cost of capital invested, which may be considered in between bank interest and the return expected if an equivalent investment was made.
3. Cost of spoilage and deterioration caused by storage and arbitrary pilferage. It is normal to use 10 % as a cost allowance for many inventory items.
4. Cost due to inflation and this may be estimated as 1 % per month of purchase cost while the item is held in inventory.

The cost of an item at the time of issue can be estimated by using the following formula:

$$I = I_c + I_F + (I_F - C_I) + k_I T C_I + k_2 C_I \qquad (5.1)$$

where

I Cost of item at time of issue;
I_c Inventorying cost;
C_I Item cost (present worth);
I_F Item cost (future worth);
k_I Inflation rate per month while the item is in stock;

k_2 Percentage of cost allowed for spoilage, deterioration, etc; and
T Time in months the item was in stock.

I_c can be estimated using the following:

$$I_c = \frac{F_S \times b}{N \times K}$$

where

F_S Floor area cost per m^2;
b bin size in m^2;
N average number of items stored in a bin; and
K reciprocal of year item spends in stock.

I_F can be calculated by using the following formula:

$$I_F = C_I(1+i)^n \tag{5.2}$$

where

i rate of interest; and
n number of interest periods

Example 1

Bin size	0.25 m^2
Number of items per bin	100
Floor area cost	\$60 per m^2
Item price	\$5.40
Inflation rate	8 % per month
Average time in inventory	4 months

$$I_c = \frac{0.25 \times 60}{100 \times 3} = \$0.05$$

$$I_F = 5.40(1+0.08)^4 = \$7.35$$

$$I = 0.05 + 7.35 + (7.35 - 5.40) + 0.08 \times 4 \times 5.40 + 0 = \$11.03$$

5.4 Maintenance Store Control Procedure

Some of the important elements of a systematic control of maintenance stores are (1) the requisition, (2) inventory record, (3) deciding what to stock, (4) order points, and (5) order quantities.

5.4.1 *Requisition*

Requisition procedures are an essential step in withdrawing material from maintenance store room. Among other uses, these procedures form the systematic basis for cost accounting and inventory control.

5.4.2 *Inventory Control*

Most control procedures make use of continuous inventory records on which receipts are added and withdrawals are subtracted. This helps in establishing the demand rate of items.

5.4.3 *Items to Be Stocked*

It is a usual practice that parts and material for routine maintenance should always be available. Parts for overhauls and non-routine maintenance should be controllable in such a manner that capital investment in spare parts is put to the best use. Classifying inventory in the least costly manner may be achieved with ABC analysis.

5.4.4 *ABC Analysis*

ABC analysis is based on Pareto's Law which states that the significant items in a group usually constitute only a small portion of the total number of items in that group. Applying this law to inventory management, it will be seen that a major proportion of the inventory value, i.e., 70–80 %, will normally comprise nearly 10 % of the number of items held in stock. This is demonstrated in Fig. 5.1.

A step-by-step procedure for constructing a Pareto's diagram is as follows:

- Select a suitable time period, usually one year, for inventory management.
- Calculate the cost of each item used in the selected period as a percentage of the total cost of inventory items.
- Rank the items in descending order of percentage of cost of that item to the total inventory cost. Starting from the items that contribute the most to the cost.
- Plot the graph with percentage of item used on the *X* axis and percentage of its cost on the *Y* axis.
- Items in class "A" are about 10–20 % of total items but account for 60–80 % of the total cost.

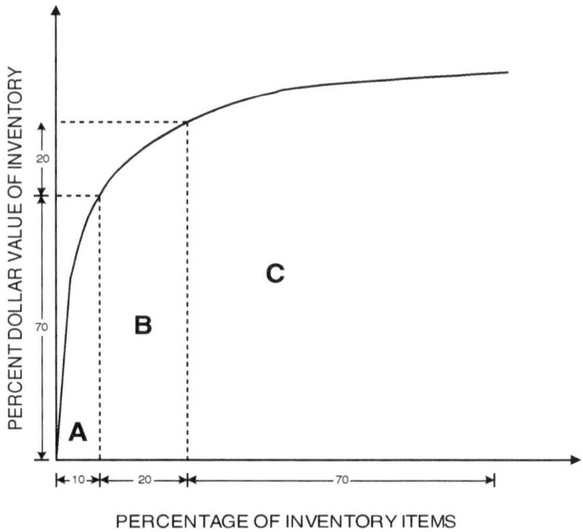

Fig. 5.1 Pareto diagram

- Items in class "B" are about 20–30 % of total items and account for 20–30 % of the total cost.
- Item in class "C" are about 60–80 % of total items but account for 10–20 % of the total cost.

It is recommended that class A items, which have high capital investment, be ordered based on calculations of most economic order quantities. Items under this class need close control. Keeping in view the high cost of these items, a minimum size of safety stock is usually maintained.

Items falling under class B may be ordered in larger quantities than class A items and similarly larger safety stocks may be maintained.

Items belonging to class C amount to 10 % of the total inventory investment and these need a minimal control and safety stocks up to 6 months may be maintained.

5.5 Inventory Systems

One practical way to establish an inventory system is to keep count of every item issued and place an order for more stocks when inventories reach a predetermined level. The order is fixed in a size which has been predetermined. Figure 5.2 illustrates such a system. In this system, the demand is known and constant. The inventory is steadily depleted until a level R_1 (reorder level) is reached. At R_1, an order of quantity Q_1 is placed. These units are assumed to arrive after a fixed and known time, usually called lead time. The demand pattern is then repeated and at point R_2, quantity Q_2 is ordered. In such a system, $R_1 = R_2$ and $Q_1 = Q_2$. This type of system is quite adequate for most of class B items and all the class C items of inventory.

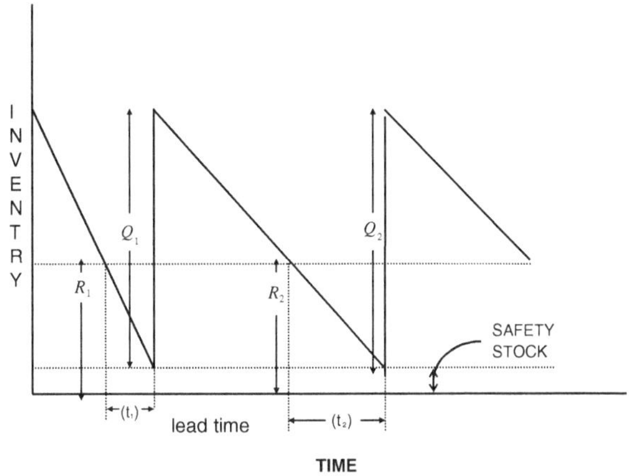

Fig. 5.2 *Q/R* inventory system

5.5.1 Economic Order Quantity (EOQ)

The main objective in inventory control is to find the minimum cost of operating an inventory system and minimize overall cost. All relevant costs that are considered to be significant are incorporated in planning the system. These include the cost of the item, procurement costs, and carrying costs. The following relationship holds

$$\text{Total Annual Cost} = \text{Cost of item} + \text{Procurement Costs} + \text{Carrying Costs}$$

Each of the costs in the equation can be stated in terms of order quantity and reorder point for a given inventory situation. A simple model in which two important cost components, procurement and carrying costs, are considered is depicted in Fig. 5.3. Because the cost of the item is taken as a constant, it is not included in this illustration.

Annual carrying costs increase with a larger value of order quantity Q. The large values of Q result in large inventory levels, and therefore, in large carrying costs. Also when Q increases, fewer orders must be placed during the year. That results in decreased procurement costs. Total costs are the sum of those two costs. This is shown as a curve in Fig. 5.3. The optimal order quantity is the point at which annual total costs are at the minimum. This point is indicated by Q^* in Fig. 5.3.

A model can be formulated for determining Q^* based on the following assumptions:

- The demand is uniform and known,
- The item cost does not vary with order size, i.e., no order discount applies for large orders,
- Complete orders are delivered at the same time.
- The lead time is known such that an order can be timed to arrive when inventory is exhausted,

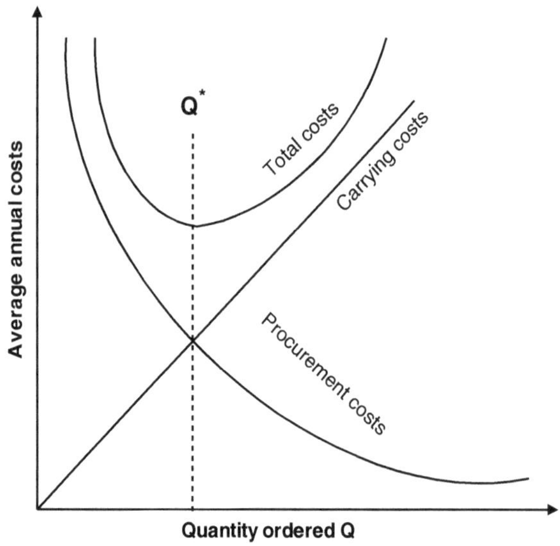

Fig. 5.3 Inventory cost model

- The cost to place and receive an order is the same regardless of order size, and
- The cost of holding inventory is a linear function of the number of items in stock.

The model minimizes the following cost function:

$$\mathrm{TC} = CD + S\frac{D}{Q} + I_{\mathrm{c}}\frac{Q}{2} \tag{5.3}$$

where

TC total annual cost;
C item cost;
D annual demand;
S ordering cost;
Q quantity ordered; and
I_{c} Inventory carrying cost.

Taking the partial derivatives of TC with respect to Q

$$\frac{\partial \mathrm{TC}}{\partial Q} = 0 + \left(-SDQ^{-2}\right) + \frac{I_c}{2} \tag{5.4}$$

and setting the first derivative equal to zero and solving for Q, we have

$$O = \frac{-SD}{Q^2} + \frac{I_c}{2} \tag{5.5}$$

$$\frac{SD}{Q^2} = \frac{I_c}{2}$$

$$Q^* = \sqrt{\frac{2DS}{I_c}}$$

Although the model is over simplified, still it is used with reasonable success by many organizations.

5.5.2 The Reorder Level

In the above model, $Q*$ is calculated under the assumption of a deterministic demand rate. The reorder level (point) corresponds to the number of items in inventory at which point an order should be triggered. It is established by taking lead time consumption into consideration, so that the new order is received when an inventory level reaches zero. In such a case, the operating doctrine would be to place an order when inventory on hand reaches the reorder level "R." If

D_{L} Demand during lead time;
d_{t} Demand rate (demand per unit time);
t_{l} Lead time; and
D_{L} $t_{\mathrm{l}} d_{\mathrm{t}}$

Example 2

Annual demand	180 units
Cost/item	\$8.00
Cost of placing order	\$9.00
Annual carrying charges	15 % of the item cost

$$\text{Then } Q^* = \sqrt{\frac{(2)(180)(9)}{0.15(8)}} = \sqrt{\frac{3240}{1.2}} \cong 52$$

In case there is a lead time of 10 days, the policy would be to order 52 when inventory on hand reaches $10\left(\frac{180}{365}\right) = 5$. If R is taken to be average demand during lead time.

5.5.3 Safety Stock

Safety stock is the average amount on hand when replenishment orders arrive. It can be thought of as the remaining inventory all year. It is usually used when the demand for items is a random variable, and therefore, inventory may reach its reorder level sooner or later than expected. Thus, the time between successive replenishments is no longer constant. There is no stock outage risk involved with demand fluctuations between the time of maximum inventory and the time inventory level reaches reorder level. The risk occurs after the reorder point has been reached. Demand during lead time may turnout to be less than, equal to, or greater than the reorder point. A safety stock may be needed to prevent stock outage during the lead time period.

An approach to determine safety stock is to use the concept of service level. A service level is the percent of times a particular item will not be outage of stock when demanded. The relationship between service level and probability of stock out age is expressed as follows:

$$\text{Service Level} = 1 - \text{Probability of stock outage}$$

In order to determine the safety stock using the service level concept, it is necessary to know the probability distribution of the demand during lead time. This can be obtained using historical data and the methodology in Appendix A, Sect. A.7. Next, it is shown how to compute safety stock using the concept of the service level by an example.

Example 3 The lead time demand of a certain type of bearing is normally distributed with a mean of 150 units and standard deviation of 5 units. The maintenance manager wants to pursue a policy, whereby the bearing is not available only 1 % of the time when demanded. Compute how much safety stock should be maintained.

Expected demand during lead time = μ = 150 units; Standard Deviation = σ = 5 units

$$\text{Let } X = \text{mean demand} + \text{safety stock} \Rightarrow \text{safety stock} = X - \mu$$

$$Z = \frac{X - \mu}{\sigma}; \text{ Where } Z \text{ can be obtained from standardized normal tables}$$

$$Z = 2.3 \text{ for } 99\,\% \text{ service level} \Rightarrow 2.33 = \frac{X - \mu}{5}$$

$$\Rightarrow \text{Safety Stock} = X - \mu = 11.65,$$

Rounding it to the nearest integer results in a safety stock of 12 units.

Thus, the restoring order should be placed when inventory is 165 = (150 + 15) units rather than 150 to achieve the desired service level.

5.5.4 *Effective Ordering Policy*

Maintenance managers must make two basic inventory policy decisions: when to reorder and how much to reorder. There are basically two reordering polices. The first one is based on a specified level of inventory (number of items) below which an inventory item is reordered. The second one is a periodic review policy that calls for ordering an item periodically rather than according to an inventory level. The amount that should be ordered is known as the order quantity. The inventory level that determines the reorder time and the reorder quantity is selected on the basis of economic considerations.

A well-known ordering policy is (s, S) policy, also known as (min, max) policy. In this policy, an order size Q_0 is placed when the inventory level reaches $s = R$ (reorder level). The order quantity is expected to arrive when the inventory level dips to the safety stock level. A good choice for Q_0 is the economic order quantity (EOQ). This policy is usually used for class A and some of class B items. For more on inventory systems see [2, 3].

An alternative policy is the two-bin policy which is usually adopted for inexpensive fast-moving items. In this policy, items are kept in the maintenance shop or factory in two bins of equal sizes. Items are drawn from one bin until it is depleted. Then, this bin is red tagged to signal that it is empty and needs to be filled and the second bin is opened for use. The sizes of the bin are the average demand in the lead time plus a safety stock.

5.6 Ordering Policies for Repairs

Statistically, the failure rate of the equipment or component varies over its life cycle. It usually depicts a definite pattern, called a bathtub curve. Figure 5.4 illustrates a typical bathtub curve.

The terms used in the figure are defined below:

1. Infant mortality: early failures due to faulty material and faulty processing.
2. Constant failure rate: random failures that have a constant rate of failure.
3. Wear out failures: failures due to aging, fatigue, etc.

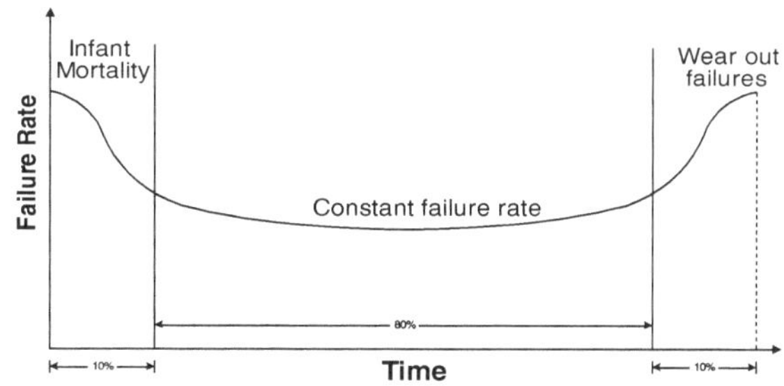

Fig. 5.4 A typical bathtub curve

The effectiveness of preventive or planned maintenance, it is generally agreed, declines through the random failure period, which the longest period in the service life of equipment. Component failure triggers equipment failure. To have the components to repair equipment during the constant failure period, component failure data and parts ordering policy must be linked.

In the earlier part of this chapter, ABC analysis of inventory according to its cost has been used for developing inventory control polices. However, spare parts must be evaluated in terms of their criticality as well, so that costly shutdowns are minimized by having critical components available when needed. A method of determining Q^* based on failure rate, along with the component criticality, is presented in the following section.

5.6.1 Demand and Failure Rate Linkage

Component failure usually triggers equipment failure. To avoid downtime due to non-availability of spares, the number of spares needed for smooth running of the equipment for a desired length of time must be estimated and made available when a demand for such parts occurs.

5.6.2 Estimating Units of Spare Needed for Replacing Failures

The two approaches for estimating demand of spares needed for desired service level are the graphical and analytical approaches. These are described below.

5.6.2.1 Graphical Approach

The graphical approach is suitable when large populations of equipment and failure data are available. This approach is explained with the help of an example in which it is assumed that pumps of a similar type fail because of failure of a particular bearing which is being considered for inventory. Table 5.1 shows running time between failures for eight pumps in use at a plant (The data are listed sequentially).

These data are plotted with run time between failures on the X axis and frequency of failures on the Y axis. The mean is 1011 h, so 50 % of the pumps are assumed to fail by this time. If management wants 80 % of the pumps operational, then enough spares of the component that causes pumps to fail should be maintained to give the desired service level. The data show 20 % of the pumps can be expected to fail by 820 h. The number of spares needed for this level of service can be calculated from the operating hours per week by providing one for each pump for each 820 h of pumping. This information is used to schedule planned or

Table 5.1 Pump failures

Pump no.	Running time	Pump no.	Running time
7	1250	6	1250
8	1450	1	1000
1	1000	2	1450
4	1500	5	700
6	1000	4	1250
2	1250	5	1000
3	700	3	700
7	600	8	600
8	500	1	1000

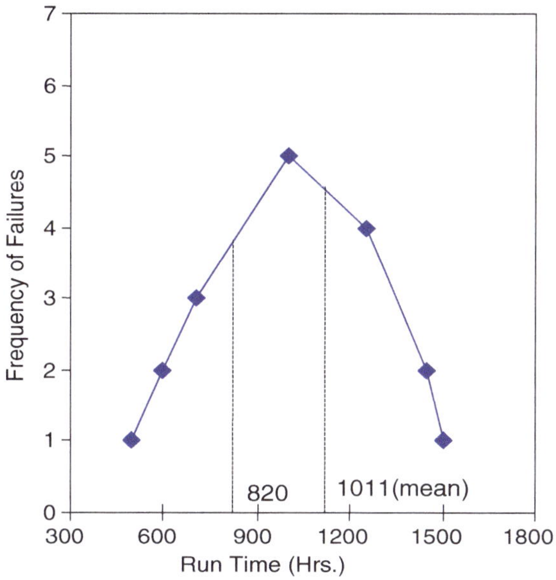

Fig. 5.5 Plot of data in Table 5.1

preventive maintenance to achieve a desired service level and have components available when needed (Fig. 5.5).

5.6.2.2 Analytical Approach

The first step is to identify the failure rate of specific parts. The following example illustrates a method for estimating the number of parts needed. The formula in Eq. (5.6) is obtained from Sheikh et al. [3].

Example 4 Consider a part which has failed 200 times for 10^6 h of a particular equipment's operation. Estimate the number of parts needed for 1-year smooth operation of the equipment with a confidence of 95 %.

Number of parts (N) needed for smooth operation can be estimated using the following formula:

$$\overset{*}{N} = \frac{t}{\bar{T}} + \sqrt{\frac{t}{\bar{T}} \times Z} \tag{5.6}$$

where

t = Operation in hours

$\bar{T} = \frac{1}{\lambda}$, where λ = failures per hour

Z = 1.65 for 95 % confidence and 2.33 for 99 % confidence

$t = (8 \times 5 \times 52) = 2080$ h

$\lambda = \frac{200}{10^6} = 0.002$ failures per hour

then $\bar{T} = \frac{1}{\lambda} = \frac{1}{0.002} = 500$

then $N = \frac{t}{\bar{T}} + \sqrt{\frac{t}{\bar{T}} Z}$ let $Z = 1.65$ for 95 % confidence

$= \frac{2080}{500} + \sqrt{\frac{2080}{500} \times 1.65}$

$= 4.16 + \sqrt{4.16 \times 1.65}$

$= 4.16 + 2.62 \cong 7$

This is estimated yearly consumption.[1]

5.7 Spare Part Classification

Spare parts need to be evaluated in terms of cost and criticality. ABC analysis is according to cost has already been discussed. Criticality can also be analyzed using the following criteria:

1. Highly critical, C_A: parts which are absolutely essential for the operation of the equipment.
2. Moderately critical, C_B: parts which if not available will have slight to moderate effect on operation of the equipment.
3. Low criticality, C_C parts which are not absolutely essential for the operation of the equipment.

[1]This is based on the assumption that time between failure is exponential and that the failure mode of the component is due to deformation or friction. For further details see Sheikh et al. [3], given at the end of this chapter.

Table 5.2 Criticality analysis

	Criticality		
Cost	C_A	C_B	C_C
A	1	1	2
B	1	2	2
C	3	3	3

(Entries indicate the strategy to be used for ordering parts in this group)

5.7.1 Spare Parts Ordering Strategies Based on Classification

Classification according to cost and critically can be grouped together as shown in Table 5.2.

5.7.2 Grouping of Spares According to Criticality and Cost

To order spares for each group, the following strategies are suggested.

Ordering Strategy 1:

For parts grouped as class 1 maintain a given quantity of items. Initial stock levels can be estimated as shown 5.6. Wherever a failure occurs procure an item (s) to replace the one consumed for repair/replacement. It is common practice to add an additional unit in the initial acquisition and use 99 % service level which correspond to $Z = 2.33$ from the standard normal table.

Ordering Strategy 2:

For parts grouped as class 2, the standard EOQ model may be used and a safety stock be maintained to offset the demand during lead time.

Ordering Strategy 3:

Items in this group are of low cost and have a varying degree of criticality. In cases where parts can be stored for a duration without any damage during storage, and parts are not readily available when needed from the market, calculate the requirement N for a considerably longer period than indicated otherwise using the approaches under 7.6.1. Alternatively, ordering strategy 2 may be used for this class as well.

5.8 Summary

Maintenance material control is just one of the ways in which total maintenance costs can be reduced. Components of maintenance stores are explained so that control functions can be effectively applied. Estimating maintenance material cost has been explained so that steps for reducing operating costs could be taken. Inventory systems, and linkage of part failure and ordering strategies are discussed.

Exercises

1. The following spares are kept in inventory at a plant. Their annual use rate and unit price are shown in the following table. Classify them as class "A," "B," or "C" items. Then suggest a policy for controlling each class.

Spare part no.	Annual use (units)	Cost per unit
S01	5000	17
S02	17,000	51
S03	6000	17
S04	4000	75
S05	11,000	200
S06	60,000	9
S07	12,000	60
S08	28,000	50
S09	4000	18
S10	3000	25

2. A taxi cab company keeps car batteries as spare part for its fleet of aging taxis. It purchases batteries at $14 each and it costs $11 to place an order. The company uses about 12,000 batteries each year at a uniform rate. The company operates 5 days a week for 52 weeks per year with the exception of six national holidays a year. The order lead time is 4 days. The holding cost is 24 % of the item cost per year. Determine the EOQ and the reorder level. What is the total inventorying and ordering cost associated with the EOQ?
3. Resolve the example given for graphical approach but with a improved service level of 95 %. Compare your results with the one in example.
4. Solve Problem 3 using analytical method. How many components are needed each year for an average running time between failures of 1011 h with 95 % confidence ($\phi = 1.65$ for 95 % confidence).

References

1. Bedworth D, Baily JE (1987) Integrated production control systems. Wiley, New York
2. Hax AC, Candea D (1984) Production and inventory control. Prentice Hall, Englewood Cliffs
3. Sheikh AK, Callom FL, Mustafa SG (1991) Strategies in spare parts management using reliability engineering approach. Eng Cost Prod Econ 21:51–57

Chapter 6
Maintenance Operations and Control

6.1 Introduction

An effective maintenance operation and control system is the backbone for sound assets management. Controlling maintenance means coordinating of demand for maintenance and available resources to achieve a desired level of effectiveness and efficiency. An effective operation and control system must have all of the following six characteristics built into it [3].

1. Maintenance demand (i.e., what work is to be done).
2. Maintenance resources (i.e., who will do the work and what material and tools are needed).
3. Procedures and means of receiving, coordinating, scheduling, dispatching, and executing the work.
4. Performance and quality standards (i.e., how long it will take to do a job and what are the acceptable specifications).
5. Feedback, monitoring, and control (i.e., the system must generate information and reports for quality, cost, and plant condition control: A mechanism for data collection integration and regular follow-up is also essential for feedback and control).
6. People or teams that are willing to change and adapt to the use of information technology (IT) to conduct the right analysis for effective maintenance decision making.

The work order system is the vehicle for planning and controlling maintenance work. It also provides the needed information for monitoring and reporting maintenance work. A clear goal and effective procedures are essential for the implementation of the work order system and control of maintenance activities.

This chapter addresses maintenance operations and control systems. It explains the necessary forms and procedures for the operations and accomplishments of maintenance work as well as the control of these operations. The concept of the

S.O. Duffuaa and A. Raouf, *Planning and Control of Maintenance Systems*,
DOI 10.1007/978-3-319-19803-3_6

maintenance control cycle is outlined in Sect. 6.2. Section 6.3 presents the work order system, and Sect. 6.4 describes some procedures and forms used for reporting and recording data other than the work order. Section 6.5 outlines the structure of the maintenance control, and Sect. 6.6 presents a host of effective engineered maintenance programs for improving the maintenance control system effectiveness. Section 6.7 contains feedback and control activities, and Sect. 6.8 proposes a host of effective engineered maintenance program. Section 6.9 outlines the role of information technology (IT) in enhancing maintenance operation and control, and Sect. 6.10 summarizes the chapter.

6.2 Maintenance Control Cycle

Maintenance can be viewed as a process and therefore a maintenance control function can apply the concepts developed in automatic process control to improve machine effectiveness. Automatic process control is a continuous cycle of the following:

- Determine the objectives and required performance.
- Sampling the output of the plant.
- Analyzing the sample.
- Applying corrective action if necessary.

Figure 6.1 shows a simple control cycle.

The concepts in Fig. 6.1 are applicable to maintenance control if used in the right framework. In maintenance, the control cycle can be defined as follows:

- The objective could be plant availability and product quality.
- Sampling output is to collect data from the work order or equipment history file.
- Analyzing the sample is to use techniques to determine whether the objective has been met. For example, does the level of quality meet customer satisfaction or does it meet specifications? Also, is equipment availability according to availability targets?

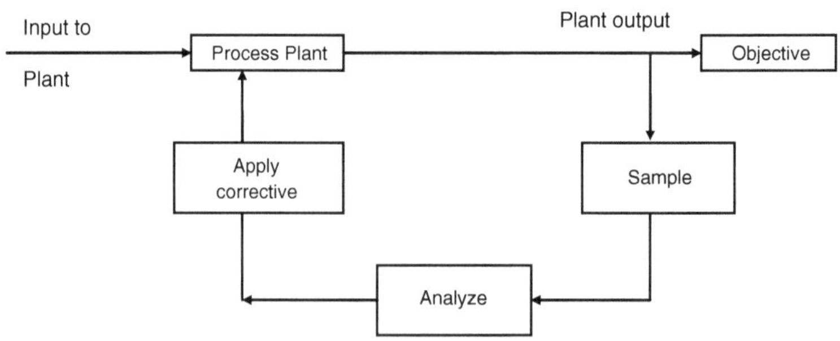

Fig. 6.1 Process control cycle

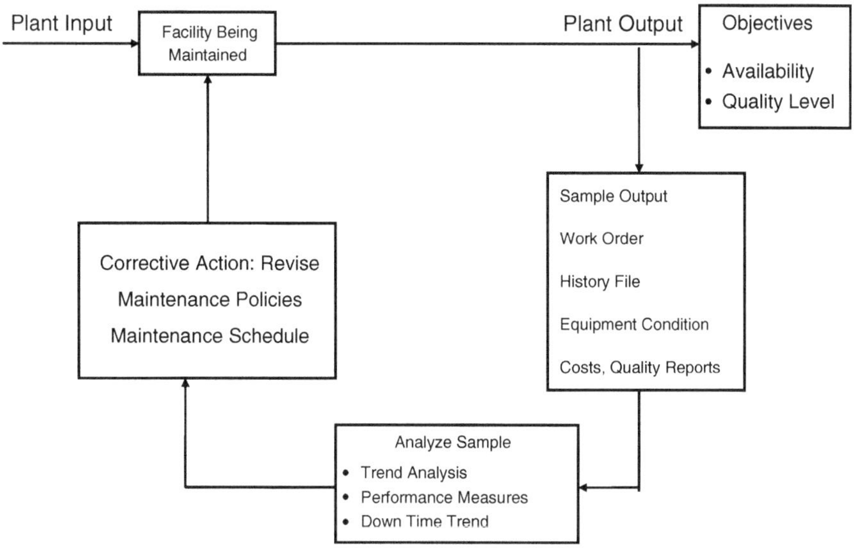

Fig. 6.2 The maintenance control cycle

- Corrective action could be revising maintenance policies, changing maintenance schedules, upgrading job specifications, training of craftsmen, and implementing new maintenance programs and strategies, if necessary.

Figure 6.2 depicts the control cycle for the maintenance process.

The application of the concepts in Fig. 6.2 requires the establishment of procedures and forms for administering maintenance work, standards for data collection and analysis, and means for effective reporting of work, equipment condition, and product quality. The latter three items are necessary for work control, cost control, and plant condition control.

6.3 Maintenance Work Order System

The first step for planning and controlling the maintenance work is through an effective work order system. The work order is a form in which written instructions are detailed for work to be carried out and it must be filled for all jobs. In industry, it is referred to by different names such as work request, work requisition, and request for service. The purpose of the work order system is to provide means for the following:

1. Requesting in writing the work to be performed by maintenance.
2. Screening the work requested by operation.
3. Assigning the best method and the most qualified crafts for the job.

4. Reducing cost through the effective utilization of resources (manpower, material).
5. Improving planning and scheduling of maintenance work.
6. Maintaining and controlling the maintenance work.
7. Improving overall maintenance through data collected from the work order and used for control and continuous improvement programs.

The administration of the work order system is the responsibility of the persons in charge of planning and scheduling. The work order must be designed with care and take into consideration two points. The first one is to include all necessary information needed to facilitate effective planning and scheduling, and the second point is to emphasize clarity and simplicity for use.

In maintenance systems, there are two types of work orders. The first type is the blanket work orders (sometimes referred to as standing work orders) used for small routine repetitive jobs, when the cost of processing an individual work order may exceed the cost of the job or when the job is a fixed routine job, such as janitor's job. In this case, the janitor does the same thing every day and his job is repetitive and preplanned. In both cases, the blanket work order provides a gross cost for generic tasks in cumulative to-date and annual costs. The maintenance planning and scheduling unit needs to define clearly what kinds of jobs are to be handled using the blanket work order.

The second type is the special work order written for all other individual jobs for which reporting all facts about the job are necessary.

6.3.1 Design of the Work Order

The work order, when used to its fullest extent, can be utilized as a work request form, a planning document, a work allocation chart, a history record (if filled), a monitoring and control tool, and as a notification of modification work completed. Therefore, care is needed in the design of the work order. The work order must contain two types of information. The first type is the information needed for planning and scheduling, and the second type is the information needed for control. Information needed for planning and scheduling includes the following:

- Inventory number, unit description, and site
- Person or department requesting the work
- Work description and time standards
- Job specification and code number
- Priority and date work is required
- Crafts required
- Spare parts and material required
- Special tools required
- Safety procedures
- Technical information (drawings and manuals).

Information needed for control includes the following:

- Actual time taken
- Cost codes for crafts
- Downtime or time work finished
- Cause and consequences of failure

A typical work order for a medium-sized company is usually filled in 3 or 4 copies, and an example of a work order is shown in Fig. 6.3. However, each work order when designed must be tailored to the specific industry where it is used. It is not very uncommon in maintenance work, while crafts are performing the job, they discover some additional work that is required and has not been included in the work plan. This could happen due to poor planning or because the damage caused by failure is larger than expected. This work should be performed, and its description is added to the work order below the dotted line in Fig. 6.3. The following is general guidelines that are applicable to most systems:

- All maintenance departments must have a work order for planning and executing their maintenance work.
- Work orders should be numbered, and at least three copies are needed for the purpose of maintenance control.
- Work order requests can be initiated by any persons in the organization and must be screened by the maintenance planner or coordinator.

6.3.2 Work Order System Flow

The work order system flow refers to the dispatching procedures and the order in which the job is processed from its initiation till its completion. In this subsection, we focus on the work order flow. The following are the sequential steps for the work order processing:

1. Upon receipt of the work request by the planner (it can be initiated via telephone, computer terminal through an enterprise resource planning (ERP), or in hard written form), it is screened and a work order is planned and completed, showing the needed information for planning, execution, and control (this chapter). Usually 3–4 copies are filled and routed in the system. If it is processed using an ERP, then the four copies are not need but routed in this fashion until completed.
2. The work order is registered in a register that lists pertinent data for each work order.
3. One copy (usually copy 1) is filed by work order number in the maintenance control department. Two copies (copy 2 and 3) are given to the concerned foreman, and one copy (copy 4) is sent to the work originator.

WORK ORDER

Work order No.	Shift	Requesting Dept ....
Date	A ☐	Cost No ....
Location	B ☐	Department ....
Equipment No..............	C ☐	Unit ....

Priority	Emergency ☐ Urgent ☐ Normal ☐ Scheduled ☐

Job must be completed uninterrupted Yes ☐ No ☐

General work description

Labor				Materials			
	Time			Parts		Price	
Trade	Est.	Act.	Detailed work description	Desc.	Part No.	Unit	Total

Job approval.. Date completed.................................

Fig. 6.3 A typical work order

4. The foreman assigns work to the appropriate craft and gives him one copy (it could be either copy 2 or 3). Let us assume it is copy 3. The craft performs the needed work and fills the necessary information about actual work done (such as actual time and actual material used) and hands the copy over to the foreman. If the company is using mobile equipment, the craft performs the job and accesses the work order through the mobile equipment and fills the necessary information about the work done.
5. The foreman verifies information and checks the quality of work and puts this verified information on copy 2. He forwards the two copies to the maintenance control. If an ERP or an e-system is used, the foreman just verifies the craft entered information and forward the work order to maintenance control.
6. The planner puts the information on copy 1 and sends copy 3 to the originator. In case an ERP or an e-system is used, the planner forwards/routes a copy to the originator.
7. The planner sends copy 2 to accounting to fill in information about costs. After that, the copy is sent to the department where the maintenance information system is maintained (it could be a unit by itself or within planning and scheduling, depending on the situation). The information in the work order is entered in equipment history file. In case an ERP or e-system is used, the work order is routed to the Accounting Department to complete cost information and the necessary information is automatically extracted from the work order to the equipment history file.
8. Copy 1 is filed in a closed work order file and kept for sometime (usually for 3–6 months) and then ultimately destroyed. In case an ERP or e-system is used, the work order is stored and utilized for generating reports and eventually archived.

Figure 6.4 shows a flowchart showing these steps. If an automated system is used, these copies can be stored as copies in the computer system and circulated via a local area network.

6.4 Record Keeping

In this section, means for recording data about maintenance work or cost are presented.

6.4.1 Job Card (Report)

In order to have the right data for work, cost, and plant condition control, it is essential to have accurate means of data collection and record keeping. Three important items need to be reported. These are (1) repair time, (2) costs, and

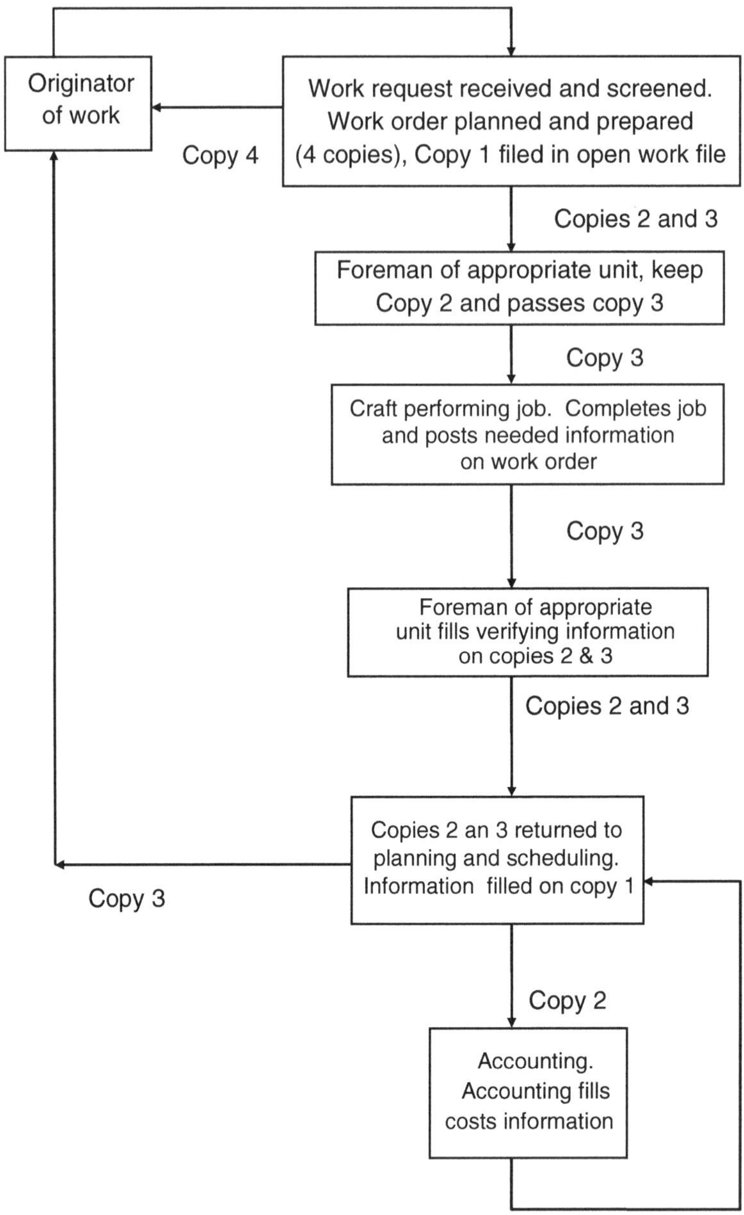

Fig. 6.4 Work order flow

(3) downtime. The job report form (job card) is a statement recording the work done and the condition of the facility. A job card can be issued to each employee participating on each job. The form can be completed manually or automated.

Job Report	Date	Report No.
Employee Name:	Craft:	Starting Time:
Facility Name	Location	Identification
Equipment defect Corrective action Spares / materials used Measurement / observations Overall equipment condition Remarks : Time taken:		

Fig. 6.5 Job card (report)

Most of the information required on the card can be extracted from the work order. In some companies, each craft records his daily work on a daily work time form where the time spent on each work order is recorded. Figures 6.5 and 6.6 provide examples of a job card and a daily work time card

6.4.2 Equipment History Record

The equipment history file is a document on which information about all work done on a particular facility/equipment is recorded. In the equipment history file, information about all repairs performed, downtime, costs of repairs, and planned maintenance specifications are recorded. It is necessary to record the following:

1. Equipment specifications and location.
2. Inspections, repairs, servicing and adjustments carried out, breakdowns and failures, and their causes and the corrective action undertaken.
3. Work done on the equipment, component repaired or replaced, condition of wear and tear, erosion, corrosion, etc.
4. Measurements or readings taken, clearance, results of tests and inspections.
5. Failure time and the time lost to carry out repairs.

There are many systems for recording and storing the information. It could be manually recorded or using a computer. The most important point is that the

Employee Name ……………………………………………………

Week Ending …………………………………… Shift ……………………………

Foreman Approval: Name ………………………… Signature ………………………

Day	Job 1		Job 2		Job 3		Job 4		Job 5		Total hours
	Number	Hours spent	Number	Hours spent	Number	Hours spent	Number	Hours spent	Number	Hours spent	
M											
T											
W											
TH											
F											
S											
SU											

M=Monday, T= Tuesday, TH= Thursday, F= Friday, S= Saturday, SU= Sunday

Fig. 6.6 Daily work time cards

Date	Maintenance performed	Downtime	Spare parts and materials	Labor	Lost Production HRS	Labor cost	Spare parts and materials cost

Fig. 6.7 Pages in equipment history file

information should be complete and recorded in an organized fashion for future access and use. A sample of a typical equipment history file is shown in Fig. 6.7.

6.5 Structure of Maintenance Control

Maintenance control comprises the following three important functions:

1. Work order coordination and planning.
2. Work order processing.
3. Information feedback and corrective action.

Figure 6.8 depicts the relationships between these functions.

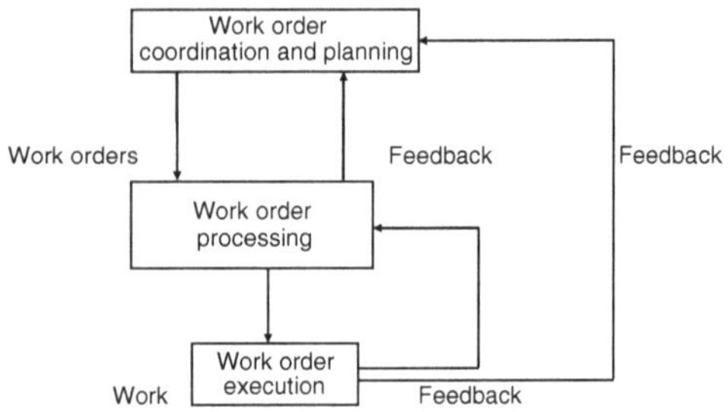

Fig. 6.8 The Structure of maintenance control

Work order coordination is concerned with satisfying maintenance demand, while requirements for production (service) and necessary capacities of maintenance resources and constraints are met. Work order processing is concerned with work order release, scheduling, and work dispatch. The feedback and control function essentially deals with information gathering and decision tasks for achieving set goals and objectives.

6.5.1 Work Order Coordination and Planning

The work order coordination function plans and coordinates the different types of maintenance requests on the basis of operational constraints, availability of resources, and priority [1, 2, 5]. This function, according to Gits [4], consists of four decision functions (Fig. 6.9).

1. Preventive maintenance (PM) planning.
2. Corrective maintenance classification.
3. Adaptive maintenance acceptance.
4. Maintenance capacity adjustment.

Preventive maintenance planning deals with the monitoring of the PM plan and generating the necessary PM work orders. The generated work orders must meet the production requirements, operational constraints, and maintenance capacity constraints.

The PM planning results in two flows of work. One type consists of fixed work orders which have to be executed by maintenance as planned. Therefore, they form an input to the scheduling directly. The other flow consists of advanceable work orders for which ultimate acceptance and execution have to be decided.

Corrective maintenance classification comprises planning and specification of the corrective maintenance work orders. In essence, this function prioritizes these work orders according to the consequences of failure for production, actual state of maintenance capacities, and operational constraints. In view of the failure, an agreement has to be reached with the affected operation on a priority system for executing this type of work.

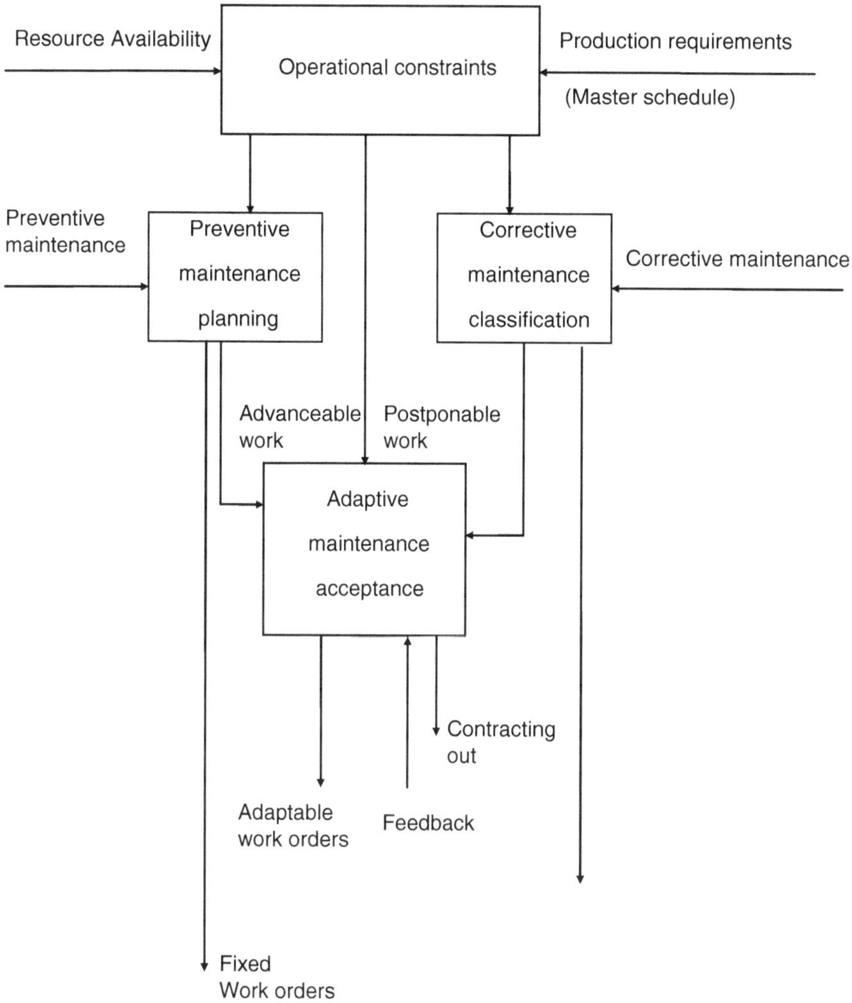

Fig. 6.9 The structure of work order coordination

Corrective maintenance classification results in two types of work flow. One flow consists of rush work orders, which have to be carried out on emergency basis, and the other flow consists of postponable work orders that can be used as a buffer-awaiting scheduling.

Adaptive maintenance acceptance attempts to smooth the work executed by the maintenance department. Fluctuations are caused by variation in the maintenance load and variability in maintenance tasks. Advanceable and postponable maintenance tasks are scheduled around rush orders when material and manpower are available. The size of this buffer should be between a minimum and a maximum level. If the buffer becomes below the minimum that provides a signal to the planner to activate PM planning to release more PM work orders. In case the maximum level is exceeded, this signals the need for overtime or contract maintenance. Control charts can be used to control the variation in work fluctuations. For more details on control charts, see Chap. 8.

Adaptive maintenance acceptance results in two types of work. The first type should be done externally, and the second type will be accomplished internally. The structure of work order coordination is shown in Fig. 6.9.

Maintenance capacity adjustment is a function needed to evaluate whether the current resources meet the maintenance demand. Issues such as the need for overtime or contract maintenance are evaluated by this function.

6.6 Work Order Processing

Work order processing focuses on realizing the work orders according to the agreements and the objectives that have to be met, taking into account the short-term production schedule. It consists of the following three control functions:

1. Work order release,
2. Work order scheduling, and
3. Work order dispatching.

The work order release function controls the amount of work orders in progress by setting free adaptable work orders from the acceptance. The major function of work order release is to decide how much work should be released per period while taking into consideration the capacity of manpower and material after considering the rush work orders.

Work order scheduling is concerned with matching resources with work and the time necessary for executing the work order. Thus, the schedule shows the resources to be employed in accomplishing the work at a given time. The final schedule provides a good idea when the work will be completed. Because of the high degree of uncertainty in predicting maintenance work, it will be impossible to adhere to the prepared schedule, and in many cases, work in progress may require rescheduling of adaptive work orders or even replanning of fixed work orders.

Work order dispatching is a control function concerned with the sequencing of work and allocating it to a specific capacity type. Dispatching refers to the procedures with which schedules are realized. It is important to have clear procedures such that everyone is informed in adequate time of their responsibility and records are maintained for both costs and information flow. Figure 6.10 outlines the structure of work order processing.

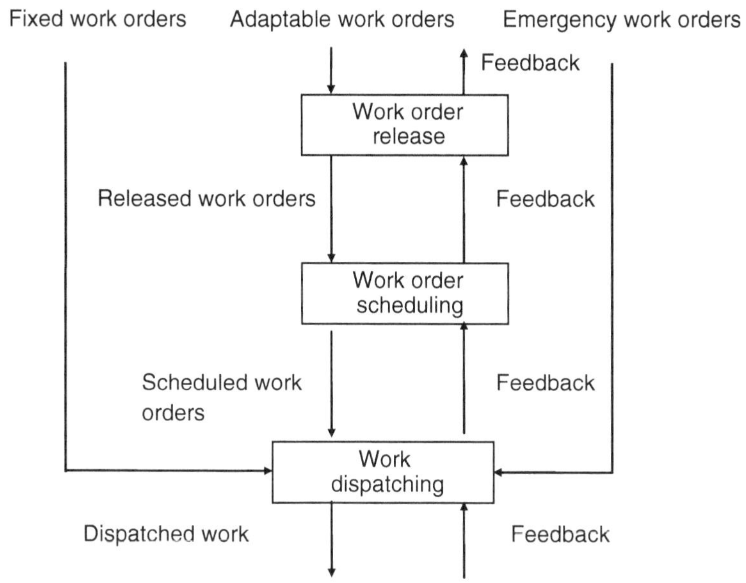

Fig. 6.10 The structure of work order processing

6.7 Information Feedback and Corrective Action

Feedback information and corrective action is concerned with the collection of data about the status of the work execution, system availability, work backlog, and quality of work performed. Then, this information is analyzed, and an appropriate course of action is formulated. This course of actions and decisions is aimed at improving the following [5–7]:

1. Work control
2. Cost control
3. Quality control
4. Plant condition control.

6.7.1 Work Control

This type of control monitors the work status and the accomplished work to investigate if the work is done according to standards (quality and time). In this type of control, it is assumed that the maintenance control system includes the standard that is assigned in advance of actual maintenance work performed. A set of reports are generated in this category of control. These include a report showing performance according to the standard by the crafts utilized for the job and their productivity. In this report, it is a good practice to indicate what proportion of maintenance work is performed using overtime. Other reports that are useful for work control are backlog, percentage of emergency maintenance to planned maintenance, and percentage of repair jobs originated as a result of PM inspection. All these reports reflect some sort of efficiency measures.

The backlog report is very essential for work control. It is a good practice to maintain a weekly backlog report by craft. The report should indicate the cause of the backlog. It is good to have a healthy backlog. The size of a healthy backlog ranges between two and four weeks of manhours of maintenance work. An excessive or too little backlog necessitates a corrective action. In case a downtrend in the backlog is identified, one of the following actions may be necessary:

1. Reduce contract maintenance.
2. Consider transfer between departments or crafts.
3. Downsize the maintenance force.

If the backlog is increasing and a clear trend is identified, a corrective action is needed which may include one of the following:

1. Increase contract maintenance.
2. Transfer between departments or crafts.
3. Schedule cost-effective overtime.
4. Increase maintenance workforce.

The total backlog should be controlled using statistical process control tools, specifically control charts (see Chap. 8). Control charts are graphs with a center line average and two control limits. The center line average for the backlog is the average of the backlog for several periods. The upper and lower control limits are functions of the variability in the backlog.

$$\text{Center Line Average} = \text{CLA} = \overline{\overline{X}} \tag{6.1}$$

$$\text{Upper Control Limit} = \text{UCL} = \overline{\overline{X}} + 3\sigma_{\overline{x}} = \overline{\overline{X}} + 3\frac{\sigma}{\sqrt{n}} \tag{6.2}$$

$$\text{Lower Control Limits} = \text{LCL} = \overline{\overline{X}} - 3\sigma_{\overline{x}} = \overline{X} - 3\frac{\sigma}{\sqrt{n}} \tag{6.3}$$

where $\overline{X}$ is the average of the backlog for several periods, σ is the standard deviation of the backlog, and n is the sample size. The following example demonstrates the use of control charts for backlog control.

In a maintenance department, the following are the weekly backlogs for the mechanical department given in terms of man-hours for the last six months (Table 6.1).

The weeks of each month are grouped as one group, and their average is computed and denoted by $\overline{X}_i, i = 1, \ldots, 6$.

Table 6.1 Weekly backlog data

Number i	Month	Week 1	Week 2	Week 3	Week 4	Weekly average $\overline{X_i}$	Weekly range R_i
1	January	180	170	200	210	190	40
2	February	200	216	250	226	223	26
3	March	180	160	150	190	170	40
4	April	170	175	160	190	174	80
5	May	210	206	226	190	208	80
6	June	195	185	175	205	190	80

There is a well-known relationship between the range of a sample from a normal distribution and the standard deviations. The random variable $W = \frac{R}{\sigma}$ where R is the range of the sample, and σ is the standard deviation of the normal distribution. W is called the relative range, and R is given as:

$$R = X_{\max} - X_{\min} \tag{6.4}$$

The parameters of W are a function of the sample size n, and the mean of W is d_2. Consequently, an estimator of σ is

$$\hat{\sigma} = R/d_2 \tag{6.5}$$

R is estimated by $\overline{R}$, the average range of several samples (groups) of the same size n. Then, an estimate of σ would be computed as

$$\hat{\sigma} = \overline{R}/d_2 \tag{6.6}$$

Table 6.2 provides values for d_2 for various sample sizes.

$$\overline{R} = \frac{R_1 + R_2 + \cdots + R_m}{m}, \quad \text{where } R_i \text{ is the range of group } i. \tag{6.7}$$

For the data in Table 6.1

$$\overline{R} = \frac{40 + 26 + 40 + 30 + 30 + 30}{6} 32.7$$

$$\hat{\sigma} = \frac{32.7}{d_2} = \frac{32.7}{2.059} = 15.9$$

$$\text{UCL} = \bar{\bar{x}} + 3\hat{\sigma} = 192.5 + 3(15.9) = 240.2$$

$$\text{LCL} = \bar{\bar{x}} - 3\hat{\sigma} = 192.5 - 3(15.9) = 144.8$$

Figure 6.11 shows the control chart for the weekly backlog.

Table 6.2 Values for d_2 for different sample size

n	2	3	4	5	6	7	8	9	10
d_2	1.128	1.693	2.059	2.325	2.534	2.704	2.843	1.970	3.078

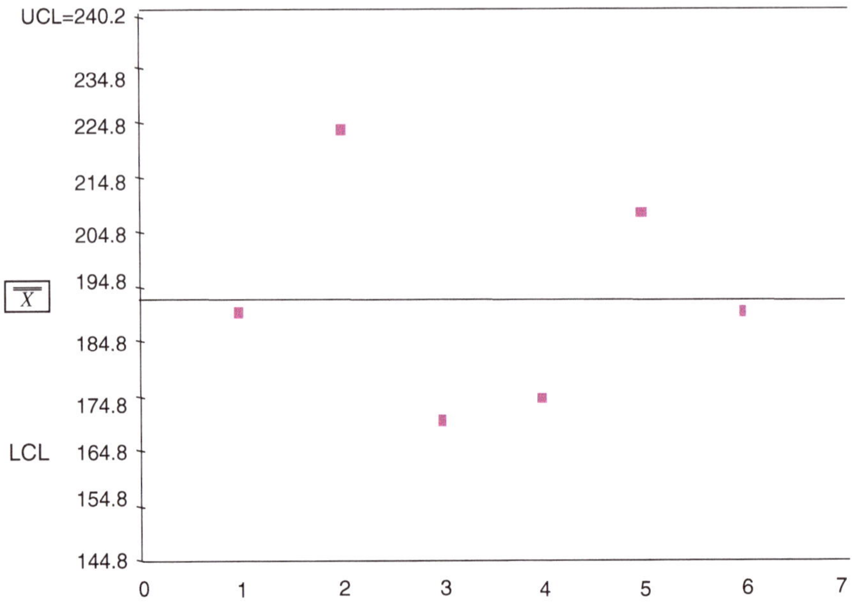

Fig. 6.11 Control chart for weekly backlog control

The backlog is in control as far as the monthly average between 144.8 and 240.2. If a monthly average is outside the limits, an action should be taken to inquire about the cause and a corrective action may be needed.

6.7.2 Cost Control

The maintenance cost consists of the following:

1. Direct maintenance cost which is the cost of labor material, spares, material, equipment, and tools.
2. Operation shutdown cost due to failure.
3. Cost of quality due to product being out of specification, as a result of machine incapability.
4. Redundancy cost due to equipment backups.
5. Equipment deterioration cost due to lack of proper maintenance.
6. Cost of over maintaining.

Almost all information about cost is available on the worker order. A summary of maintenance costs by work must be issued monthly. This is utilized to control maintenance costs and develop costs of manufactured products.

The cost reports will indicate which are the most needed cost reduction programs. Cost reduction should be a continuous effort in any sound maintenance program. The areas where cost reduction programs can be launched are as follows:

1. Considering the use of alternatives maintenance materials.
2. Modifying inspection procedures.
3. Revising maintenance procedures, particularly making adjustments in size of crew and methods.
4. Redesigning material handling procedures and the workshop layout.

6.7.3 Quality Control

Maintenance has a direct link to the quality of products. Well-maintained equipment produces less scrap than poorly maintained equipment. It is also well documented that the condition of the machine affects its process capability. Current research has established the link between maintenance and quality. For example, the capability of a machine tool that is in top conditions may have a tolerance distribution that has 99 % of the product within specifications. However, a worn machine will have more chatter and vibration. The tolerance distribution of a worn machine will be wider and show a larger variance, and a lower percentage of the production will be within specification.

A monthly report on the percentage of repeat jobs and product rejects may help identify which machine requires an investigation to determine causes of quality problems. Once the machines are investigated, a corrective course of action will be taken to remedy problems. The action may entail a modification in the current maintenance policy and training of crafts. (For more details, see Chap. 8.)

6.7.4 Plant Condition Control

Plant condition control requires an effective system for recording failures and repairs for critical and major equipment in the plant. This information is usually obtained from the work order and equipment history file. The records in the equipment history file include the time of failure, the nature of failure, the repairs undertaken, total downtime, and machines and spares used.

A monthly maintenance report should include downtime of critical and major equipment and their availability. If downtime is excessive or the equipment availability and readiness is low, a corrective action must be taken to minimize the occurrence of failure. The corrective action may require establishment of a reliability improvement program or a planned maintenance program, or both.

6.8 Effective Engineered Maintenance Program

In this section, six engineered maintenance programs are outlined. These programs offer sound courses of action that can be adopted in a maintenance control cycle. The objective of these programs is to improve plant availability, reduce cost, and improve equipment reliability and product quality. These programs are listed below:

1. Planned maintenance
2. Emergency maintenance
3. Reliability improvement
4. Cost reduction
5. Training and employee motivation
6. Equipment management program.

6.8.1 Planned Maintenance

Planned maintenance is an integrated effort to convert most of the maintenance work into scheduled maintenance. The planned maintenance is the work that is identified through preventive and predictive maintenance. It includes inspection and servicing of jobs that are performed at specified recurring intervals. It also includes condition-based maintenance.

In planned maintenance, all activities are preplanned. This includes material planning and stocking. Material planning permits more reliable scheduling in addition to cost savings in material delivery and ordering. Also, the jobs will be scheduled at times that do not disrupt deliveries or production schedules. The savings from introducing planned maintenance are significant in terms of reduction in downtime and material costs. Planned maintenance offers a sound approach to improve maintenance and enables it to meet set objectives.

6.8.2 Emergency Maintenance

Emergency maintenance refers to any job that should be started in the same day. Emergency maintenance by its nature allows very little lead time for planning. The amount of emergency maintenance must be minimized, and it should not exceed 10 % of the total maintenance work. The maintenance department must have a clear policy for handling emergency maintenance. Following are two approaches for handling emergency maintenance. These are as follows:

1. Introduce the emergency maintenance into the regular schedule and then pick up the backlog with either overtime, temporary workers, or contract maintenance. It is an accepted practice in industry to allow 10–15 % of load capacity for emergency work.

2. Estimate the amount of emergency maintenance and assign skilled dedicated crafts for the emergency work order.

In most plants, the first approach is adopted because it is expected to result in increased workforce utilization; however, the second approach offers the ability to respond fast. It is necessary, as soon as the work has started, and it is possible to estimate the amount of repair needed for the emergency job, to plan the remainder of the job based on the information available.

6.8.3 Reliability Improvement

A reliability engineering program offers a sound alternative for improving the maintenance function. It can be used as an option to improve maintenance performance. Critical and major equipment history files must be maintained, and estimates for mean time between failures (MTBF) must be calculated. The frequency of emergency maintenance is a function of the failure rate of this equipment. It can be estimated for a period of operations lasting n hours, and there will be n/MTBF emergency maintenance actions. The longer the MTBF, the lower the number of emergency maintenance incidents.

Another approach that enhances equipment reliability and optimizes maintenance operations is through a reliability-centered maintenance (RCM) program. In RCM, the maintenance program is developed on the basis of the concept of restoring equipment function rather than bringing the equipment to an ideal condition. RCM has been applied successfully in the commercial airline industry, nuclear reactors, and power plants. A systematic approach for developing an RCM program is described in Chap. 13.

6.8.4 Equipment Management Program

Total productive maintenance (TPM) (see Chap. 13) is a Japanese philosophy that focuses on equipment management in order to improve product quality. Its aim is to reduce six equipment losses to improve overall equipment effectiveness (OEE). The six major causes of equipment losses, according to Nakajima, are as follows:

1. Failure,
2. Setup and adjustments,
3. Idling and minor stoppage,
4. Reduced speed,
5. Process defects,
6. Reduced yield.

Also, the equipment management program focuses on developing a sound TPM program for each equipment and providing a good approach for improving maintenance performance.

6.8.5 Cost Reduction

Maintenance can contribute in reducing product cost through a continuous effort of cost reduction in maintenance operations. Cost reduction in maintenance can be obtained by applying engineering method approaches. These approaches study how the work is being performed with the aim of developing a better way to perform maintenance. Engineering methods have well-defined steps for examining the maintenance work in order to simplify and eliminate unnecessary steps. This reduction and simplification of work results in cost savings. In the effort aiming at reducing costs, the following should be considered:

1. Alternative material and spare parts.
2. Alternative method for inspection and overhaul.
3. Alternative equipment and tools.
4. Alternative procedures for planning and scheduling.
5. Alternative job time standards.

The engineering method approach offers valuable means for improving maintenance and controlling costs.

6.8.6 Training and Employee Motivation

Production maintenance (act of performing maintenance) depends to a large extent on the skilled crafts. Many of the maintenance ineffectiveness can be traced to the lack of skilled technical staff. Therefore, it is necessary to have an ongoing on-the-job-Training (OJT) program to help assure that employees are equipped with the necessary skills for effective maintenance. The skills include judgment, skills in communicating and reading technical information, and in some cases skills in multiple fields (multi-skilling.)

The training program should include off-the-job and on-the-job training. The maintenance department must have a yearly training program to upgrade and update the knowledge of its staff. The training should include a modern technique program that periodically brings the latest maintenance techniques to all concerned.

The training program must be coupled with a motivational program. Employee motivation can be achieved by incentive program that rewards the productive crafts and encourages continuous improvement.

6.9 Role of Information Technology (IT)

In the previous sections of this chapter, the elements and forms for designing an effective maintenance operations and control are presented. This includes procedures, forms, measures, and possible improvement actions. In the past, the elements of the maintenance operation and control are carried manually or semi-manually; however, in the current age of information technology, it would be a waste to continue to operate manually. Information technology enhances communication, data sharing, and analysis, improves maintenance decision making, and plants reliabilities.

IT enables designing work orders and work order flows (Figs. 6.3 and 6.4), equipment history files (Fig. 6.5), and other maintenance control structures on an ERP to enable speedy communication and timely decision making. The knowledge about assets operation and control and IT with trained people enhances reliability and predictability and business performance.

6.10 Summary

Maintenance control systems play a key role in having an effective maintenance program. The concepts of automatic process control that include process objectives, sampling, sample analysis, and applying corrective actions are defined and used to establish an effective maintenance control system. Then, the structure of maintenance control is explained in detail. It consists of work order coordination, work order processing, and feedback information and corrective action. The steps for effective maintenance control are as follows:

1. Define objectives and goals in terms of quality, availability, and efficiency.
2. Utilize IT to avail information timely and aid in effective maintenance decision making.
3. Coordinate and plan work orders.
4. Process work orders.
5. Collect information from work orders and history files and compile reports on efficiency, availability, and quality.
6. Examine the deviation from established objectives and goals.
7. If deviation exists, take corrective action otherwise improve goals.

The six engineered maintenance programs described in Sect. 6.6 offer ways and means for formulating a corrective action if needed. They also provide ways for improving the current maintenance status.

Exercises

1. Describe the maintenance control cycle.
2. List three objectives for maintenance control.
3. What is the purpose of the work order? Describe its flow in the maintenance system.
4. What are the methods and actions you may need to employ per backlog control?
5. Design a work order for a medium-sized organization.
6. Visit three organizations in your area and collect a sample of their work order. For each organization, point out the deficiencies in their work order and suggest improvement.
7. What are the three control functions in work order processing?
8. What are the possible courses of actions available for improving maintenance performance to achieve stated objectives?
9. Propose two schemes for handling emergency maintenance.
10. Reliability-centered maintenance (RCM) and total productive maintenance (TPM) are programs that can help improve maintenance control performance. Describe how each one of these programs can help in having a better control over maintenance performance.

References

1. Al-Sultan KS, Duffuaa SO (1995) Maintenance control via mathematical programming. J Qual Maintenance Eng (JQME) 1(3):36–46
2. Ben Daya M, Duffuaa SO (1995) Maintenance and quality: the missing link. J Qual Maintenance Eng (JQME) 1(1):20–26
3. Clifton RH (1974) Principles of planned maintenance. Edward Arnold Publishers, Ltd.
4. Gits CW (1994) Structuring maintenance control. Int J Oper Prod Manage (IJOPM), Spec Issue Maintenance Manage 14(7):5–17
5. Kelly A (1984) Maintenance planning and control. Butterworth, London
6. Nakajima S (1988) Introduction to TPM. Productivity Press Inc., Norwalk
7. Neibel BW (1984) Engineering maintenance management, 2nd edn. Marcel and Dekker, New York

Chapter 7
Maintenance Planning and Scheduling

7.1 Introduction

Planning is the process by which the elements required to perform a task are determined in advance of the job's start time. Scheduling deals with assigning the jobs to be accomplished at a specific time. It is clear that good planning is a prerequisite for sound scheduling; however, for successful planning, feedback from scheduling is necessary. This is the reason why in many maintenance organization, they are performed by the same person or unit. Planning and scheduling in maintenance are different than production in the following aspects [1–2, 7–8]:

- The demand for maintenance work has more variability than production and the arrival of the demand is random in nature.
- Maintenance jobs have more variability between them, even the same type of jobs differs greatly in content. This made it harder to develop job standards in maintenance than production, see Chap. 4. Reliable job standards are necessary for sound planning and scheduling.
- Maintenance planning requires the coordination between many departments in an organization such as material, operations, and engineering, and in many situations, it is a major cause of delays and bottlenecks.

The above reasons necessitate a different treatment for maintenance planning and scheduling. Planning and scheduling are the most important aspects of sound maintenance management.

Effective planning and scheduling contribute significantly to the following:

- Reduced maintenance costs. Studies done by several researchers including Alcan;
- General motors have shown that a clear link exists between planned maintenance and reduced costs;

S.O. Duffuaa and A. Raouf, *Planning and Control of Maintenance Systems*,
DOI 10.1007/978-3-319-19803-3_7

- Improved utilization of maintenance work force by reducing delays and interruptions. It also provides a good means for improving coordination and making supervision easier; and
- Improved quality of maintenance work by adopting the best methods, procedures, and assigning the most qualified crafts for the job.

The principal objectives of planning and scheduling include the following:

- Minimizing the idle time of maintenance forces;
- Maximizing the efficient use of work time, material, and equipment; and
- Maintaining the operating equipment at a level that is responsive for the need of production in terms of delivery schedule and quality. In fact, almost all maintenance should be planned and scheduled. Only emergency work is performed without advance planning. Emergency work should be planned as it is progressing.

For planning purposes, maintenance work can be classified into the following five categories:

1. *Routine and preventive maintenance* include periodic maintenance such as lubricating machines, inspections, and minor repetitive jobs. This type of work is planned and scheduled in advance.
2. *Corrective maintenance* involves the determination of the causes of repeated breakdowns and eliminating the cause by design modification.
3. *Emergency or breakdown maintenance* is the process of repairing as soon as possible following a reported failure. The reporting is usually done through phone and followed by a confirming work order. Maintenance schedules are interrupted due to repair emergency breakdowns.
4. *Scheduled overhaul which involves the shutdown of the plant* is planned and organized such that the plant shutdown is minimized.
5. *Scheduled overhaul*, repairs or building of equipment which does not fall under the above categories.

An essential part of planning and scheduling is to be able to forecast future work and to balance the work load among the above five categories. The maintenance management system should aim to have over 90 % of the maintenance work to be planned and scheduled in order to reap the benefits of planning and scheduling. Ref [9] provides a good introduction to forecasting.

This chapter will present the ingredients and the techniques of effective planning and scheduling. Section 7.2 presents the elements of planning, and Sect. 7.3 presents the ingredients of sound scheduling. Section 7.4 presents the importance of job priorities for scheduling. Section 7.5 outlines the techniques for scheduling, and Sect. 7.6 presents the turnaround maintenance. Section 7.7 summarizes and concludes this chapter.

7.2 Planning

Planning in the context of maintenance means the process by which all the elements required to perform a task is determined and prepared prior to starting the execution of the job. The planning process comprises all the functions related to the preparation of the work order, bill of material, purchase requisition, necessary drawings, labor planning sheet, job standards, and all the data needed prior to scheduling and releasing the work order. Therefore, an effective planning procedure should include the following steps:

1. Determine job content (may require site visits).
2. Develop work plan. This entails the sequence of activities in the job and establishing the best methods and procedures to accomplish the job.
3. Establish crew size for the job.
4. Plan and order parts and material.
5. Check whether special equipment and tools are needed and obtain them.
6. Assign workers with the appropriate craft skill.
7. Review safety procedures.
8. Set priorities (emergency, urgent, routine, and scheduled) for all maintenance work.
9. Assign cost accounts.
10. Fill the work order.
11. Review backlog and develop plans for controlling it.
12. Predict the maintenance load using an effective forecasting technique.

The maintenance work order usually does not provide enough space to perform the details of planning for extensive repairs, overhauls, or large maintenance projects. In such cases where the maintenance job (project) is large and requires more than 20 h, it is useful to fill a maintenance planning sheet. Such sheets were found to be useful in planning the maintenance of freight cars in railways, when the cars arrive for their six-month-scheduled preventive maintenance. In the maintenance planning sheet, the work is broken down into elements. For each element, the crew size and the standard times are determined. Then, the content of the planning sheet is transferred in one or several work orders. In filling the planning sheet or the work order, the planner must utilize all the expertise available in the maintenance department. Thus, consultations with supervisors, foremen, plant engineers, and crafts should be available and very well coordinated. Figure 7.1 is a sample of a maintenance planning sheet. Therefore, the planning and scheduling job require a person with the following qualifications:

1. Full familiarity with production methods used through the plant;
2. Sufficient experience to enable him to estimate labor, material, and equipment needed to fill the work order;
3. Excellent communication skills;
4. Familiarity with planning and scheduling tools; and
5. Preferable with some technical education.

Sheet......... of	Filled by..	Date ....
Equipment #	Approval	Priority Normal □ Scheduled □

No.	Completion date	Work Order #	Unit	Work description	Crafts	Estimated Time
....	....	....	....	....	....	....
....	....	....	....	....	....	....
....	....	....	....	....	....	....
....	....	....	....	....	....	....
....	....	....	....	....	....	....
....	....	....	....	....	....	....
....	....	....	....	....	....	....
....	....	....	....	....	....	....
....	....	....	....	....	....	....
....	....	....	....	....	....	....
....	....	....	....	....	....	....
....	....	....	....	....	....	....
....	....	....	....	....	....	....
....	....	....	....	....	....	....
....	....	....	....	....	....	....
....	....	....	....	....	....	....
....	....	....	....	....	....	....
....	....	....	....	....	....	....
....	....	....	....	....	....	....
....	....	....	....	....	....	....
....	....	....	....	....	....	....
....	....	....	....	....	....	....
....	....	....	....	....	....	....

Fig. 7.1 Maintenance planning sheet

The planner office should be centrally located and the office organization depends on organization size.

The planning process can be divided into three basic levels depending on the planning horizon:

1. Long-range planning (covers a period of five years).
2. Medium-range planning (1 month to one year plans)
3. Short-range planning (daily and weekly plans).

For long- and medium-range planning, the planner needs to utilize the following methods:

1. Sound forecasting techniques to estimate the maintenance load.
2. Reliable job standard times to estimate manpower requirements.
3. Aggregate planning tools such as linear programming to determine resource requirements.

The above techniques have been presented in Chap. 5 and are necessary for specifying the required manning levels. The long-range plan covers a period of 3–5 years and sets plans for future activities and long-range improvements.

The medium-range plan covers a period of one month to a year. The plan will specify how the maintenance force will operate and provide details for major overhauls, construction jobs, preventive maintenance plans, plant shutdowns, and vacation planning. This plan balances the need for manpower over the period covered and estimates required spare parts and material acquisition.

Short-range planning concerns periods of one day to one week. It focuses on the determination of all elements required to perform industrial tasks in advance. The planner has to go through steps 1 through 12 given at the beginning of this section.

7.3 Scheduling

Maintenance scheduling is the process by which the jobs are matched with resources (crafts) and sequenced to be executed at certain points in time. A reliable schedule must take into consideration the following:

1. A job priority ranking that reflects the urgency and the criticality of the job.
2. Whether all the materials needed for the work order are in the plant. (If not, the work order should not be scheduled.)
3. The production master schedule and close coordination with operation.
3. Realistic estimates and what is likely to happen rather than what the scheduler desires.
4. Flexibility should be built in the schedule. The scheduler must realize that flexibility is needed especially in maintenance. The schedule is often revised and updated.

The maintenance schedule can be prepared in three levels depending on the horizon of the schedule. The levels are (1) long-range or master schedule to cover a period of 3 months to one year; (2) weekly schedule; it is the maintenance work that covers a week; and (3) the daily schedule covering the work to be completed each day.

The long-range schedule is based on existing maintenance work orders including blanket work orders, backlog, preventive maintenance, and anticipated emergency maintenance. It should balance long-term demand for maintenance work with available manpower. Based on the long-term schedule, requirements for spare parts and material could be identified and ordered in advance. The long-range schedule is usually subjected to revisions and updating to reflect changes in plans and realized maintenance work.

The weekly maintenance schedule is generated from the long-range schedule and takes account of current operation schedules and economic consideration.

The weekly schedule should allow for about 10–15 % of the work force to be available for emergency work. The planner should provide the schedule for the current week and the following one, taking into consideration the available backlog. The work orders that are scheduled for the current week are sequenced based on priority. Critical path analysis and integer programming are techniques that can be used to generate a schedule. In most small- and medium-sized companies, scheduling is performed based on heuristic rules and experience.

The daily schedule is generated from the weekly schedule and is usually prepared the day before. This schedule is frequently interrupted to perform emergency maintenance. The established priorities are used to schedule the jobs. In some organizations, the schedule is handed to the area foreman and he is given the freedom to assign the work to his crafts with the condition that he has to accomplish jobs according to the established priority.

7.3.1 Elements of Sound Scheduling

Planning the maintenance work is a prerequisite for sound scheduling. In all types of maintenance work, the following are the necessary requirements for effective scheduling.

1. Written work orders that are derived from a well-conceived planning process. The work orders should explain precisely the work to be done, the methods to be followed, the crafts needed, spare parts needed, and priority.
2. Time standards that are based on work measurement techniques as explained in Chap. 4.
3. Information about craft availability for each shift.
4. Stocks of spare parts and information on restocking.
5. Information on the availability of special equipment and tools necessary for maintenance work.
6. Access to the plant production schedule and knowledge about when the facilities may be available for service without interrupting the production schedule.
7. Well-defined priorities for the maintenance work. These priorities must be developed through close coordination between maintenance and production.
8. Information about jobs already scheduled that are behind schedule (backlogs).

The scheduling procedure should include the following steps as outlined by Hartman:

1. Sort backlog work orders by crafts,
2. Arrange orders by priority,
3. Compile a list of completed and carry over jobs,
4. Consider job duration, location, travel distance, and possibility of combining jobs in the same area,
5. Schedule multi-craft jobs to start at the beginning of every shift,
6. Issue a daily schedule (except for project and construction work), and
7. Have a supervisor check work assignments (perform dispatching).

The above elements provide the scheduler with the requirements and the procedure for developing a maintenance schedule.

For large jobs or maintenance projects, especially shutdown maintenance jobs, the scheduler can use available quantitative techniques for generating the schedule and balancing manpower requirements. These techniques include critical path analysis (CPM), Program Evaluation and Review Techniques (PERT), Integer Programming, and Stochastic Programming. These methods are presented in the later sections of this chapter. Next, the role of priority in maintenance scheduling is presented together with a methodology for developing the jobs priorities.

7.4 Maintenance Job Priority System

The maintenance job priority system has a tremendous impact on maintenance scheduling. Priorities are established to ensure that the most critical and needed work is scheduled first. The development of a priority system should be well coordinated with operations staff, who commonly assigns a higher priority to maintenance work than warranted. This tendency puts stress on the maintenance resources and might lead to less than optimal utilization of resources. Also, the priority system should be dynamic and must be updated periodically to reflect changes in operation or maintenance strategies. Priority systems typically include three to ten levels of priority. Most organizations adopt four or three level of priorities. Table 7.1 provides the classification of the priority level and candidate jobs to be in each class.

7.5 Scheduling Techniques

The ultimate objective of scheduling is to construct a time chart showing the start and finish time for each job (activity), the inter-dependencies among job, and the critical jobs that need a special attention and effective monitoring.

Table 7.1 Priorities of maintenance work

Priority		Time frame work should start	Type of work
Code	Name		
1	Emergency	Work should start immediately	Work that has an immediate effect on safety, environment, quality will shut down the operation
2	Urgent	Work should start within 24 h	Work that is likely to have an impact on safety, environment, quality, or shutdown the operation
3	Normal	Work should start within 48 h	Work that is likely to impact the production within a week
4	Scheduled	As scheduled	Preventive maintenance and routine. All programmed work
5	Postponable	Work should start when resources are available or at shutdown period	Work that does not have an immediate impact on safety, health, environment, or production operations

In the past, the scheduling of jobs in a project was done based on heuristic techniques, and the first-known scheduling tool was the Gantt chart developed by Henry L. Gantt during World War II. The Gantt chart is a bar chart that specifies the start and finish time for each activity on a horizontal timescale. Its principal disadvantage is that it does not show interdependencies among different jobs. The Gantt chart can be modified to show interdependencies by noting milestones on each job timeline. The milestones indicate key time periods in the duration of each job. Solid lines connect interrelationships among milestones. The milestones thus indicate the interdependencies between jobs. Obvious milestones for any job are the starting time for the job and the required completion point. Other important milestones are significant points within a job, such as the point at which the start of other jobs is possible. Figures 7.2 and 7.3 show examples of these charts.

Project planning and scheduling techniques have evolved over time and two important analytic techniques are widely used for planning and scheduling. These are the critical path method (CPM) and project evaluation and review technique (PERT). CPM was first developed by E.I. due Pont de Nemours and Company as an application to the construction projects and later was extended by Mauchly Associates. PERT was developed by the US Navy for scheduling research and development projects. Other mathematical programming models that are useful for job scheduling are integer programming and stochastic programming.

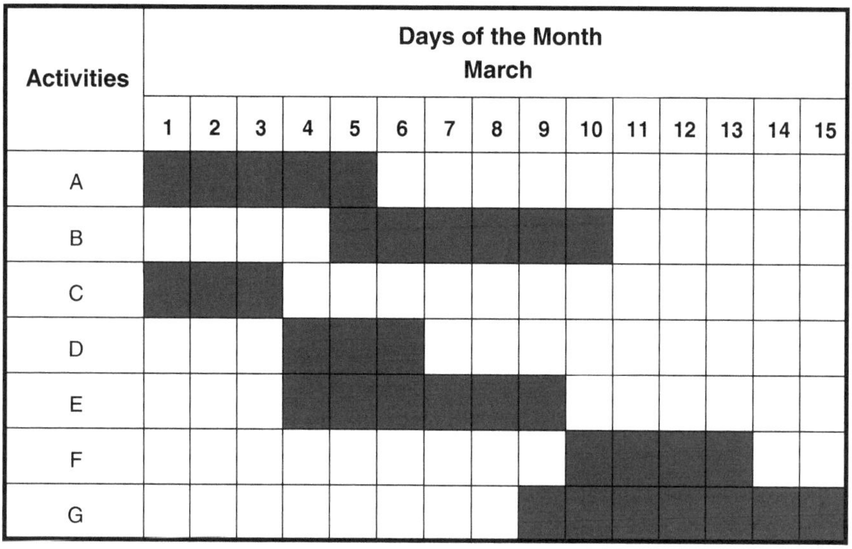

Fig. 7.2 Gantt chart

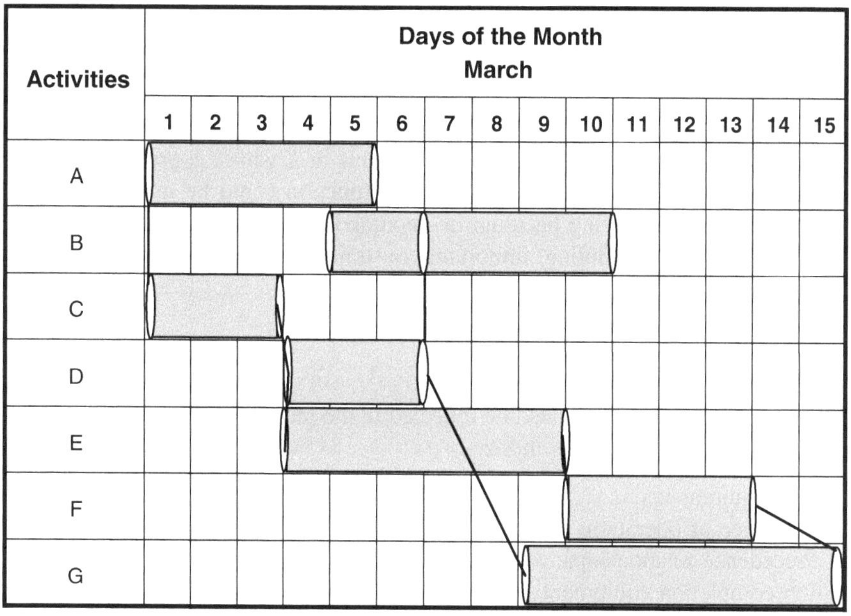

Fig. 7.3 Gantt chart with milestones

7.5.1 *Mathematical Programming Approaches for Maintenance Scheduling*

Mathematical programming is a term that has been coined by Dorfman and refers to a host of operations research models. In a general form, the mathematical programming formulation is as follows:

$$\text{Minimize}\, f(x)$$
$$\text{Subject to}\; x \in X$$

where X is the feasible set of points defined by a set of constraints and $\in$ is the belonging notation. In other words, mathematical programming finds a feasible $x \in X$ (that satisfies the constraints) and optimizes the given objective function. The objective function and constraints could take any form such as linear, quadratic, and/or a general nonlinear functional form depending on the problem. Depending on the form of the set X and the objective function, one of the following mathematical programming models could result:

1. Network programming.
2. Linear programming.
3. Quadratic programming.
4. Nonlinear programming.
5. Stochastic programming.

In the maintenance scheduling frame of reference, the set X represents all feasible schedule. A schedule is composed of points in X whose coordinates are time for each job to be performed. The objective function could be minimizing work force idle time, minimizing backlog, or maximizing manpower utilization.

In maintenance scheduling, important constraints are the following:

1. Manpower (skills) availability;
2. Equipment and tools; and
3. Spare parts availability.

The above constraints could also be reflected in the physical and planning elements of maintenance jobs, such as these:

1. Job requirements;
2. Sequence of operations for each job;
3. Precedence relationships; and
4. Job completion equipment.

Moreover, some of the constraints or the objective could be either deterministic or stochastic in nature. Some of the above-mentioned mathematical programming models have been applied to maintenance scheduling more extensively than others. These are the network models and integer programming. The network model includes both the CPM and PERT. In the next sections, the network models and the integer programming application to maintenance scheduling will be presented in detail and the use of stochastic programming will be highlighted. For more on mathematical programming and its application in maintenance planning and scheduling see [3–6, 10–12].

7.5.2 Critical Path Method

The application of CPM and PERT will yield a schedule that specifies the start and finish time for each job. A prerequisite for the application of both methods is the representation of the project as a network diagram, which shows the interdependencies and precedence relationships among the activities of the project. PERT and CPM are not well suited for day-to-day independent small jobs scheduling in a maintenance department; however, they are very useful in planning and scheduling large jobs (20 man hours or more) that consist of many activities. Examples of such jobs are overhauls of machines, and coordinating plant shutdown and turnaround maintenance activities. The following are the steps for applying CPM:

1. Develop the project network diagram. This is a representation of the job in terms of its activities, showing all precedence relationships between activities.
2. Perform the CPM calculation to identify the critical jobs. (critical jobs are those jobs on the critical path and non-critical jobs are those jobs with float).
3. Perform project crashing to (determine minimum times for each job) reduce project duration and investigate the cost trade-offs.
5. Level of the resources in order to have uniform manpower requirements to minimize hiring, firing, or overtime requirements.

7.5.3 Network Diagram Representation

The network diagram allows the representation of a project (a large job) to be seen as an integrated system. Interaction and precedence relationships can be seen easily and be evaluated in terms of their impact on other jobs. In this type of network, an arrow is used to represent an activity (a job) and a node represents an event. An event is a point in time that signifies the start or completion of an activity.

The direction of the arrow indicates the direction of progress in the project. Activities originating from a certain node cannot start until the activities terminating at this node have been completed. Figure 7.4 shows a typical representation of an activity (i, j), its start point is node i, termination node is j, and duration is as shown next to the arrow. Figure 7.4b shows that activity (1, 2) must be completed prior to starting activities (2, 3) and (2, 4).

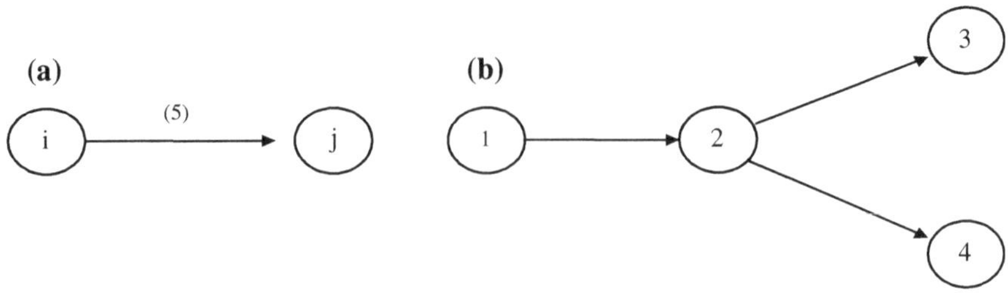

Fig. 7.4 Activity representation

The following are the rules for constructing the project network.

1. Each activity is represented by one and only one arrow in the network.
2. No two activities can be represented by the same starting and ending nodes. In certain circumstances, this may require the creation of dummy activities to adhere to this rule. A dummy activity is an activity with zero duration.
3. To develop a correct network representation of the project that shows the right precedence relationship, the following questions must be answered when an activity is added to the network.

   3.1. What activities must be completed immediately before this activity?
   3.2. What activities must start immediately after this activity?
   3.3. What activities must occur simultaneously with this activity?

For more on project network representation refer to Taha [12].

The project network representation will be demonstrated by an example from maintenance. Table 7.2 shows the data for overhauling a bearing in a train cargo carriage. The data show the normal, crash duration, and the normal crash costs precedence relationships for each activity. The term crash time refers to the minimum time the job can be accomplished in, beyond which no further reduction in the job duration can be achieved. At this duration, any increase in the resources for this job will increase the cost without reducing the duration.

Following the rules specified above for constructing the network diagram, Fig. 7.5 is obtained as follows: we notice that job A is the only job that starts at time 0 at node 1. Job A can be followed by jobs B, C, and D; however, job E cannot start until job B is completed. Therefore, the jobs B, C, and D originate from node 2. Job B is followed by job E emanating from node 3. Job F originates from node 4 after job C has been completed and so on. Following the rules above and the precedence relationship, we obtain the network shown in Fig. 7.5.

Table 7.2 Normal and crash CPM data for bearing overhaul

Job (activity)	Description	Time (minutes)		Costs ($)		Immediate precedence relationship
		Normal	Crash	Normal	Crash	
A	Dismantling	50	30	100	150	0
B	Repair of bolster pockets	67	50	120	150	A
C	Repair side frame rotation stop legs	90	60	150	200	A
D	Check friction blocks and all spring	35	25	50	75	A
E	Repairing of bolster rotation stop gibs	35	25	140	170	B
F	Repair side frame column wear plates	55	40	100	130	C
G	Repair bolster pivot	210	150	250	300	E
H	Assemble	65	45	120	150	D, F and G
I	Painting	40	30	80	100	H

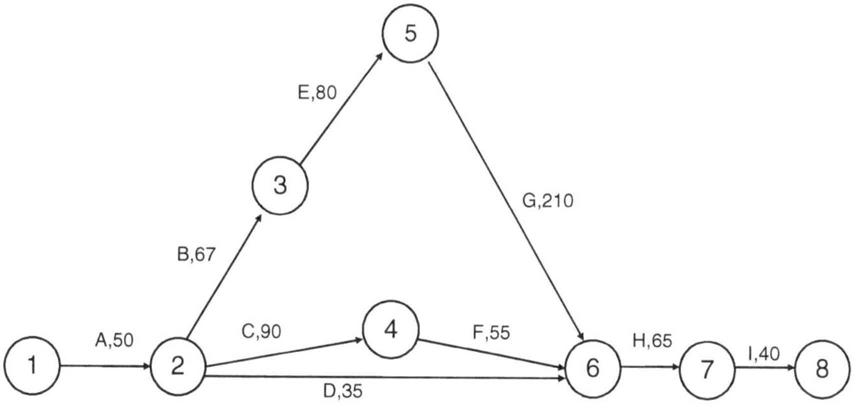

Fig. 7.5 Network diagram for bearing overhaul data in Table 7.2

7.5.4 Critical Path Calculation

The application of CPM and PERT will distinguish critical jobs from non-critical jobs. Critical jobs are those jobs which, if a delay in their starting time occurs, will cause a delay in the completion of the whole project. A job is non-critical if the difference between its earliest starting time and its latest completion time is more

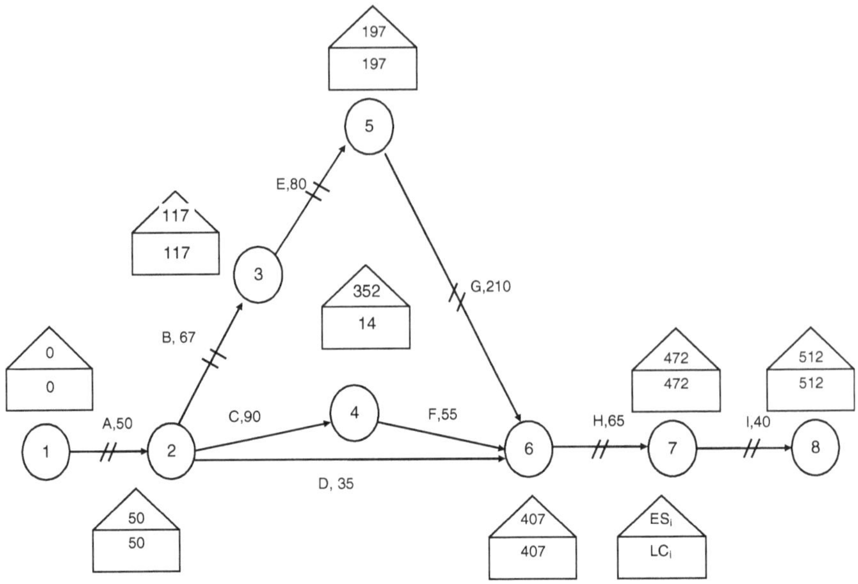

Fig. 7.6 Results of critical path calculation

than its duration. These jobs permit flexibility in choosing the start time and may be worked intermittently within the allowable time frame. This is termed slack or float. The critical jobs must be monitored carefully and adhere to their specified schedules; however, non-critical jobs can be used for leveling the resources due to the available slack.

A critical path is a chain of critical jobs (activities) that connect the starting node (event) to the last (ending) node of the network diagram. The critical path calculation includes two phases. The first phase is the forward path where calculation starts from the beginning node and progresses forward to the last node of the project. At each node, a number is computed representing the earliest time for the occurrence of the event (node). This number, which is a time, is the earliest starting time (ES) for all jobs (activities) emanating from this node. The second phase, the backward path, starts calculation from the end node and moves backward to the starting node. The numbers computed in this path represent the latest occurrence time of this event without delaying the project. This number represents the latest completion (LC) time for all activities entering this node. In Fig. 7.6, the earliest starting times are shown in squares □ and the latest completion times are shown in triangles Δ next to each node. The formula for calculating these times is given below. The variables used in the calculation are defined as follows:

D_{ij} The duration of job (i, j)
ES_i The earliest start time of all activities emanating from node i
ES_0 0 (Earliest starting time at the beginning node (node 0))
LC_i Latest completion time for all activities entering node i

LC_n ES_n, where n is the ending node
ES_j $\max\{ES_i + D_{ij}\}$ for all activities (i, j) in the network diagram
LC_i $\min\{LC_j – D_{ij}\}$ for all activities (i, j) in the diagram

The forward path for the bearing overhaul network diagram is as follows:

$ES_1 = 0$
$ES_2 = ES_1 + D_{12} = 0 + 50 = 50$
$ES_3 = ES_2 + D_{23} = 50 + 67 = 117$
$ES_4 = ES_2 + D_{24} = 50 + 90 = 140$
$ES_5 = ES_3 + D_{35} = 117 + 80 = 197$
$DS_6 = \text{Max } [ES_2 + D_{26}, ES4 + D_{46}, DS_5 + D_{5,6}]\text{Max } [50 + 35, 140 + 55,197 + 210] = 407$
$ES_7 = ES_6 + D_{67} = 407 + 65 = 472$
$ES_8 = ES_7 + D_{78} = 472 + 40 = 512$

The backward path is as follows:

$LC_8 = 512$
$LC_7 = LC_8 - D_{78} = 512 - 40 = 472$
$LC_6 = LC_7 - D_{67} = 472 - 65 = 407$
$LC_5 = LC_6 - D_{56} = 407 - 210 = 197$
$LC_4 = LC_6 - D_{46} = 407 - 55 = 352$
$LC_3 = LC_3 - D_{35} = 197 - 80 = 117$
$LC_2 = \text{Min } [LC_3 – D_{23}, LC_4 – D_{24}, LC_6 – D_{26}] = \text{Min } [117 - 67, 352 - 90, 407 - 35] = 50$
$LC_1 = LC_2 - D_{12} = 50 - 50 = 0$

The results of the forward and backward paths are shown in Fig. 7.6.

The critical path can be identified from these calculations. An activity (i, j) lies on the critical path if it satisfies the following conditions:

$ES_i = LC_i$
$ES_j = LC_j$
$ES_j - ES_i = LC_j - LC_i = D_{ij}$

In other words, an activity (i, j) is on the critical path if there is no slack or float time between the earliest start (completion) and the latest start (completion) time of the activity. The critical path for the bearing overhaul example is shown by the arrows that have two strokes on them. It consists of activities (jobs) (1,2), (2,3), (3,5), (5,6), (6,7), and (7,8). The non-critical activities are (2,5), (5,6), and (2,6), i.e., jobs C, D, and F.

The final result of the CPM calculation should provide a schedule for all the activities that level the resources. The critical activities must be scheduled at their earliest starting time since these jobs have no slack. The non-critical activities provide flexibility in scheduling and leveling the resources. Each non-critical activity has two types of float. They are the total float (TF_{ij}) and free float (FF_{ij}), and they are defined as follows:

TF_{ij} $LC_j - ES_i - D_{ij}$
FF_{ij} $ES_j - ES_i - D_{ij}$

The free float assumes that all activities start as early as possible. The critical path calculations with the float for the non-critical activities are shown in Table 7.3. This table includes all information needed to develop a time chart showing the schedules for all the activities.

The construction of the time chart should be made taking into consideration the available resources and must take full advantage of the CPM calculation. In some circumstances, it might not be possible to schedule many activities simultaneously because of personnel and equipment limitations. The total float for non-critical activities can be used to level the resources and minimize the maximum resource requirement. These activities can be shifted backward and forward between maximum allowable limits and scheduled at an appropriate time that levels the resources and keeps a steady work force and equipment. For more on project scheduling, see Phillips and Taha [12, 13].

In addition to resource leveling, CPM involves project crashing. In project crashing, the duration of one or more critical activities is shortened in an optimal fashion and a curve is prepared to show the trade-off between time and cost. This will enable management to evaluate project duration with the resulting cost. Network programming can be used to perform crashing in an optimal fashion.

7.5.5 Program Evaluation Review Techniques (PERT)

The CPM does not incorporate uncertainty, nor direct cost considerations in project scheduling. PERT addresses the issue of uncertainty by assuming that the time estimate for activity *(i, j)* duration is based on three different values, as below:

O_{ij} optimistic time, which is the time required if execution goes extremely well,
P_{ij} pessimistic time, which is the time required under the worst conditions
m_{ij} most likely time, which is the time required under normal condition

Assuming a beta distribution, the mean (D) and the variance (V) for the activity duration time is given by

$$\overline{D}_{ij} = \frac{O_{ij} + P_{ij} + 4m_{ij}}{6}$$

$$V_{ij} = \left(\frac{b_{ij} - a_{ij}}{6}\right)^2$$

If the activities on the network are assumed independent, then X_i be the occurrence time of node event i. X_i is a random variable and it is the sum of the duration of the

Table 7.3 Critical path method calculations

Job (activity) (i, j)	Duration (Dij) (minutes)	Earliest		Latest		Total float LCj	Free float TFij	Critical activities
		Start ESi	Completion ECij	Start LSij	Completion LCij			
(1,2)	50	0	50	0	50	0	0	*
(2,3)	67	50	117	50	117	0	0	*
(2,4)	90	50	140	262	352	212	0	
(2,6)	35	50	85	372	407	322	322	
(3,5)	80	117	197	117	197	0	0	*
(4,6)	55	140	195	352	402	212	212	
(5,6)	210	197	407	197	407	0	0	*
(6,7)	65	407	473	407	472	0	0	*
(7,8)	40	472	512	472	512	0	0	*

LS_{ij} = Latest start time for activity $(i, j) = LC_j$
ES_{ij} = Earliest completion time for activity $(i, j) = ES_i + D_{ij}$
* = critical activities
Note: It is not always true $FF_{ij} = TF_{ij}$

activities on the path (P) from the starting node to node i. If there is more than one path, the situation is rather involved; however, the path with the largest uncertainty as reflected by its variance is taken in this case.

X_I is the sum of independent random variables, and by the central limit theorem, the distribution of X_i is approximately normal with mean $E(X_i)$ and variance Var (X_i), where

$$E(Xi) = \sum_{(i,j) \in R} \overline{D}_{ij}$$

$$\mathrm{Var}(X_i) = \sum_{(i,j) \in P} V_{ij}$$

Using the above assumption, we can calculate the probability with which a project can be completed. If the schedule of occurrence of node i is TS_i, then

$$\begin{aligned} \Pr[X_i \le TS_i] &= P\left[\frac{X_i - E(X_i)}{\sqrt{\mathrm{Var}(X_i)}} \le \frac{ST_i - E(X_i)}{\sqrt{\mathrm{Var}(X_i)}}\right] \\ &= \Pr[Z \le Z_i] = \Phi(Z_i) \end{aligned}$$

Φ is the distribution function of the standard normal distribution.

7.5.6 Integer and Stochastic Programming Approaches for Maintenance Scheduling

The integer programming model can be stated as

$$\mathrm{Min} \sum_{j=1}^{n_1} c_j x_j \tag{7.1}$$

subject to

$$\mathrm{A}x \le \mathrm{b}$$

$$x_j = 0 \text{ or } 1$$

where $A = (a_{ij})$ is $m_1 \times n_1$; $b = (b_j)$ is $m_1 \times 1$; and $C = (c_j)$ is $1 \times n_1$ vector. All A, C, and b are known constants. The matrix represents the coefficients of the constraints such as required hours of each skill for jobs, b represents available skills, and c_j represents cost values or penalties for unfinished jobs. Then, x_j values provide the times at which jobs are scheduled.

Integer programming models are appropriate if all the details about the maintenance system are known in advance (i.e., the maintenance system is deterministic), which in turn makes c_j's, A, and b as known constants. However, maintenance systems are stochastic in nature, since the maintenance system has many stochastic elements that are not known in advance and can only be predicted at best. This shows that the maintenance scheduling problem must have both deterministic and stochastic parameters, and stochastic constraints that result from uncertain anticipated emergency.

There are generally two ways of approaching this problem. One of them is anticipative namely to try to expect what will happen in the future and plan for it. The other one is adaptive, which is to say wait until something is needed and then act accordingly. Each approach has its advantages and disadvantages.

Stochastic programming with resource models are those which combine both anticipative and adaptive strategies in one model in an attempt to strike a balance between the long-term advantage of the anticipative approach and the short-term advantage of the adaptive approach. This translates into trade-off between capital investment and short-term operation, between the cost of maintaining a slowly varying work force size to the cost of hiring, lay-off, and overtime of a fast-changing work force.

Stochastic programming with resource (SR) is a formulation that handles uncertain future events. The stochastic programming model with simple resource (SR) is stated as follows:

$$\text{Max} \quad cx + E[\inf q^{+}y^{+} + q^{-}y^{-} \mid y^{+} - y^{-} = R - Tx, y^{+}, y^{-} \geq 0] \tag{7.2}$$

subject to

$$\mathrm{Ax} \leq \mathrm{b}$$

$$\mathrm{x} \geq 0$$

where R is a vector of random variable with known distribution function, and E is the expectation operator; A *is* $m_1 \times n_1$, b *is* $m_1 \times 1$, c *is* $1 \times n_1$, $q+$ is $1 \times m_2$, $q-$ is $1 \times m_2$, and T is $m_2 \times n_1$ are the given matrices. Another way of stating the SR model is as follows:

$$\text{Max } cx + E[\inf q^{+}y^{+} + q^{-}y^{-}] \tag{7.3}$$

subject to

$$\mathrm{Ax} \leq \mathrm{b}$$

$$\mathrm{Tx} + \mathrm{y}^{+} - \mathrm{y}^{-} = \mathrm{R}$$

$$\mathrm{x} \geq 0, \ \mathrm{y}^{+}, \mathrm{y}^{-} \geq 0$$

The deterministic constraints, $Ax \leq b$, represent the constraints for jobs on hand as explained in the case of the integer programming model and the second set of constraints represents the stochastic component of the model. The above formulation for the integer and stochastic programming is given in general forms. For more details about both models, see Duffuaa and Al-Sultan [5].

In order to implement the mathematical programming approaches, the following steps must be undertaken for preparing the model and solving it.

1. Decide on an objective to be used. In the example below, the selected objective is to minimize the number of unfinished jobs. The jobs are prioritized according to importance. This has been decided by the criticality of the job to the plant. In the stochastic case, we need to estimate probability of job arrivals at each horizon.
2. Formulate model constraints. These include constraints on manpower availability, skill per hour balances, and selection constraints. The last set of constraints deals with which job to select and at which hour for scheduling. In the stochastic case, we may need to have reserved manpower for anticipated jobs. This must be reflected in the constraints and the objective.
3. If the maintenance manager (coordinator) is interested only in scheduling jobs on hand, his appropriate model is the deterministic integer program. In case he has elements of uncertainty and he would like to reserve manpower for anticipated jobs, his appropriate approach is the stochastic model. The complexity of these models depends on the plant type and size.
4. Depending on the formulated model, packages are available for solving mathematical programming models. Examples of these are Linear Interactive and Discrete Optimizer (LINDO), General Integer and Nonlinear Optimizer (GINO), and Optimization Software Library (OSL).
5. Results of the models will yield implementable schedules. However, to investigate the sensitivity of the results to parameters of the model, techniques such as sensitivity analysis can be invoked.

7.5.6.1 Integer Programming Model Formulation

In this section, an integer programming model is formulated for scheduling maintenance personnel and is demonstrated by a simple example. The model is the one given in [10] developed by Roberts and Escudero.

A few assumptions are made to formulate this model, which are as follows: Job duration times are known constants, spare parts are available and their delivery times are negligible, and equipment and tools are available all the time. The limiting factor in this problem is manpower.

The model uses the following notations:

i subscript, for the ith skill; $i = 1, 2, \ldots, S$

j subscript, for the jth job

k subscript, for the kth hour clock time

l subscript, for the lth starting time for a job

t subscript, for the tth hour from the start of a job, incremental time
h_{ijt} hours required for the ith skill on the jth job, at the tth hour from the start of the job
H horizon of scheduling hours
n number of known jobs to be scheduled (deterministic part)
N total number of jobs (both known and expected)
S total number of skill types available
S_{ik} skill hours available, for ith skill, kth hour clock time
Y_{ik} binary variable for the jth job, kth hour clock time, lth starting time for the job; if $Y_{jkl} = 1$, the job is scheduled. If $Y_{jkl} = 0$, *the job is not scheduled*
C_{jkl} values associated with Y_{jkl}; taken to be one in this model
θ_j job duration, stated as the number of clock hours for job j
$\{I_j\}$ index set for skill types required for job j
$\{I_k\}$ index set of skill types that are used in hour k
$\{J_{ik}\}$ index set for jobs that use skill i in hour k, for all t and l subscripts
n_j duration of job j in hours

The objective function of the model is to maximize the number of scheduled jobs in the horizon H. This implicitly maximizes personnel utilization.

Job-hour balances

This set of inequalities gives the job description in terms of the hours and skill required for each job, hour-by-hour. These constraints will determine the time the job must start in order to be completed before the end of the planning horizon. For example, if the planning horizon is 8 h and a job requires 4 h, then this job may begin in hours 1, 2, 3, 4, or 5, but not in hours 6, 7, or 8 because as a 4-h job, it will end beyond the planning horizon.

$$\sum_{i\in\{I_j\}}\sum_{t=1}^{\theta_j} h_{ijt}Y_{ill} \leq \sum_{i\in\{I_j\}}\sum_{t=1}^{\theta_j} h_{ijt}, \quad j = 1,\ldots,n \text{ and } l = 1,\ldots,\mathrm{H} - \theta_j + 1$$

Skill-hour balances

These sets of constraints provide hour-by-hour (to each skill) requirements for the jobs, which must be less than the skill availability for each hour.

$$\sum_{j\in\{J_{ik}\}} h_{ijt}y_{ill} \leq S_{ik}, \quad i = 1,2,\ldots,S, \quad k = 1,2,\ldots,H$$

Selection equations

These equations assume that each job can be run once and only once.

$$\sum_{l=1}^{H-\theta_j+1} Z_{jl} = 1 \quad j = 1, 2, \ldots, n$$
$$Y_{ill} - Z_{jl} = 0 \quad j = 1, 2, \ldots, n, \quad l = 1, \ldots, H - \theta_j + 1$$

The model above can be stated in a compact form as follows:

$$\text{Max} \quad \sum_j \sum_k \sum_l C_{jk} l z_{jl}$$

subject to

Job-hour balances

$$\sum_{i \in \{l_j\}} \sum_{t=1}^{\theta_j} h_{ijt} Y_{ill} \le \sum_{i \in \{l_j\}} \sum_{t=1}^{\theta_j} h_{ijt}, \quad j = 1, \ldots, n \text{ and } l = 1, \ldots, \mathrm{H} - \theta_j + 1$$

Skill-hour balances

$$\sum_{j \in \{j_{ik}\}} h_{ijt} y_{ill} \le S_{ik}, \quad i = 1, 2, \ldots, S, \quad k = 1, 2, \ldots, H$$

Selection equations

$$\sum_{l=1}^{H-\theta_j+1} Z_{jl} = 1 \quad j = 1, 2, \ldots, n$$

In order to demonstrate the utility of the proposed model, the following example taken from Roberts and Escudero is considered. The horizon taken is an 8-h shift and the block for each job shows the skill requirements for that job. For example, job 1 requires 1 plumber in the first hour and 1 electrician in the second and third hours. M represents mechanic, E electrician, and *P* plumber. Jobs 1 through 8 are the jobs on hand at the beginning of the horizon and are shown in Table 7.4.

The maintenance department has 2 plumbers, 1 electrician, and 1 mechanic. The full deterministic formulation is given in the appendix of this chapter. The problem was solved by LINDO—an acronym for Linear Interactive Discrete Optimizer package and the optimal solution obtained as shown in Table 7.5.

Horizon

The objective function value of the optimal schedule is 8. Clearly, the above schedule is only one of several alternative optimal schedules.

The integer programming formulation can be extended to a stochastic programming formulation by adding the stochastic component to the model and reserving needed manpower for anticipated jobs. This can be accomplished by adding new variables defined as reserved manpower for anticipated jobs and this

Table 7.4 Jobs on hand at the beginning of the horizon

Job 1	P	E	E		
Job 2	P	E	M	M	M
Job 3	E	P			
Job 4	M	M	P		
Job 5	E	P			
Job 6	P	P	M		
Job 7	E	M			
Job 8	M	E	P		

variable can be determined in an optimal fashion by penalizing excess and unavailability of manpower in case a job is realized. This has been developed and demonstrated by an example in [5].

7.6 Turnaround Maintenance

Turnaround (TA) maintenance is a periodic maintenance in which plants are shutdown to allow for inspections, repairs, replacements, and overhauls that can be carried out only when the assets (plant facilities) are taken out of service. During turnaround maintenance, the following types of work will be performed:

1. Work on equipment which cannot be done unless the whole plant is shutdown;
2. Work which can be done while equipment is in operation but requires a lengthy period of maintenance work and a large number of maintenance personnel; and

Table 7.5 Optimal schedule for the deterministic formulation

Job	1	2	3	4	5	6	7	8	
1		P	E	E					
2				P	E	M	M	M	
3							E	P	
4					M	M	P		
5								E	P
6	P	P	M						
7	E	M							
8	M	E	P						

3. Defects that are pointed out during operation, but could not be repaired, will be maintained during turnaround period.

The turnaround concept originated in process industries, and it is a big event for all petrochemical industries.

The overall objective of turnaround (TA) maintenance is to make all the equipment to operate properly and safely in order to maximize production capacity. Specifically, the following are the objectives of TA maintenance:

1. Expand or modify the assets to realize projected revenues by increasing production capacity, minimizing operation cost, and reducing downtime.
2. Minimize risks to employees in the immediate area surrounding the operating equipment.

3. Achieve budget figures and ensure the forecasted economic life of the assets is achieved.
4. Modify operating equipment to cope with legal requirements and or obligations such as environmental regulation.

TA maintenance is an event that requires all the elements of effective maintenance industry: organization, planning, scheduling, reporting, costing, and continuous improvement. In the following subsections, we briefly describe these elements.

7.6.1 Turnaround Maintenance Planning

The planning for TA should start 6–8 months prior to starting the execution of the actual work. The planning process is derived by the objectives stated above, and for each objective, a corresponding operational objective is formulated and coupled with a set of maintenance tasks to achieve TA maintenance objectives.

The first objective is realized through the following two operational objectives, which are as follows:

- Prevention of production unavailability, and
- Reduction of downtime.

The above operational objectives are achieved through the following tasks:

1. Debottlenecking of equipment and systems to ensure quantity and quality of products.
2. Replacement of components and material that can cause undesirable performance. In petrochemical plants, the corresponding action would be to replace or replenish catalysts and chemicals.
3. Renew or modify into permanent condition standards, equipment, and systems which are in temporary service.
4. Overhaul and/or replace parts of specified equipment and machinery whose mechanical life can be predicted, especially those machines who's overhauling can only be executed under close supervision of a specialist.

The second objective is related to risk and safety which can be realized through the following operational objective:

- Preservation and improvement of safety and pollution.

The above operational objective is achieved through the following tasks:

1. Ensure emergency shutdown and/or safety interlock systems are in top condition.
2. Ensure failure and relief systems, especially those restoring pressure relief values, are in top operating conditions.
3. Inspect and test the pollution and safety control devices and equipment.

The third objective focuses on achieving budget and cost figures and can be realized by the following operational objective:

- Prediction of future maintenance and imminent component failure or time-dependent deterioration.

The above operational objectives can be achieved through the following tasks:

1. Inspect specified equipment and collect technical data to predict imminent component failure.
2. Develop effective condition-based maintenance program and use techniques that predict the current conditions of the equipment.
3. Use the latest analytical tools for prediction such as trend analysis and time series analysis.

The fourth objective deals with compliance with legal requirements, such as environmental regulations and international quality standards such as ISO 9000 and ISO 14000. This objective can be realized through the following operational goal:

- Tie the maintenance procedures and practices with legal requirements, environmental regulations, and quality standards.

The above operational objective can be achieved through the following tasks:

1. Review the latest legal requirements, new environmental regulations, and company plans for adopting new quality assurance standards.
2. Identify equipment which when operated violates standards.
3. Develop and implement modification to equipment and procedures that facilitate compliance with requirements, regulations, and quality standards.

The above set of tasks will determine the volume of maintenance work during the TA maintenance period. However, care should be exercised to determine the essential work in order not to over spend in this period.

Table 7.6 provides an example of the type of work estimated for a petrochemical plant for stationary and running equipment. This table is used to estimate the needed manpower and the cost, and an example of such estimation is shown in Table 7.7.

7.6.2 Turnaround Schedule

The schedule for TA maintenance should consider the following:

1. Legal or contractual limitations.
2. Operation schedule.
3. Nature of the process.
4. Enough lead time for preparing an overall plan, material purchase, and assuming manpower availability.
5. Operation schedule of other related industries.

Table 7.6 Guideline for TA maintenance planning

Category	Work item	Work volume
Stationary equipment		
A, D	Inspection and cleaning of reactors, vessels, and columns	100 % of equipment debottlenecking, 100 % of equipment for scale deposition, corrosion, and other mechanical defects anticipated (3)
A	Inspection and cleaning of storage tanks	Only selected tanks requiring works inside the tanks
A, D	Inspection and cleaning of heat exchangers	100 % of equipment for debottlenecking, 100 % of equipment for selected reboilers, preheaters, and coolers, for scale deposition, corrosion, and other mechanical defects anticipated, except equipment with spare ones (1) (3)
B	Inspection of relief system	100 % of equipment showing malfunctions, selected equipment of extremely important safety valves Selected equipment of breather valves and other pressure relief system (1)
A, D	Inspection, cleaning and repair of piping, valves and fittings	100 % of piping under temporary services, 100 % of piping for scale deposition, corrosion, and other mechanical defects anticipated selected piping, special materials, and/or lining applied (3)
A, D	Inspection and cleaning of unit equipment	100 % of or selected unit equipment for, scale deposition, corrosion and other mechanical defects anticipated, except equipment with spare ones (1)(2)(3)
Running equipment		
A, D	Inspection of centrifugal compressors	100 % of centrifugal compressors, blowers, and turbines for partial or complete overhauls except equipment with spare ones (1)(2)(3)
A	Inspection of reciprocating compressors	100 % of reciprocating compressors without spare ones, or selected equipment with spare ones, for partial or complete overhauls considering lifetime of parts (1)(2)(3)
A, D	Inspection of pumps	Selected pumps, specialty and manufactured, for partial or complete overhauls (1)(2)(3)

Then, an overall schedule for the TA maintenance should be developed. Figure 7.3 provides an example of such a schedule. Then, a detailed schedule for large jobs using critical path analysis should be prepared for planning, monitoring, and control. A bar chart must result for each size of job.

A two-shift per day schedule must be extracted from the master schedule. For each shift, a meeting should be held to review progress and plan for appropriate actions, if the daily plan is behind schedule.

Table 7.7 Work and manpower requirements

	Manpower			Cost (SR)		
	Plant	Contractor	Total	Manpower	Material	Total
1. Mechanical	23	155	178	1,250,000	550,000	1,800,000
2. Instrument/ Analyzer	9	19	28	385,000	43,000	428,000
3. Electrical	6	12	18	127,320	28,000	155,320
4. Computer	2	–	2	–		–
Grand total	40	186	226	1,762,320	621,000	2,383,320

7.6.3 Turnaround Maintenance Reporting

The TA maintenance reporting consists of the following:

1. Daily progress report, used mostly for monitoring and control.
2. TA maintenance report detailing all the jobs carried out in different areas and giving facts and figures of the TA maintenance.
3. Turnaround costing report. It consists of material cost, manpower cost, and contract services.

The reports in 2 and 3 are used for improvement purposes.

7.7 Summary

In this chapter, maintenance planning and scheduling have been addressed. The planning process is a prerequisite for scheduling. The elements of effective planning have been explained followed by description of scheduling. Techniques for scheduling are presented. These include CPM, PERT, and Integer Programming. CPM is very useful for project scheduling or when the maintenance job is large and has many activities. Integer Programming is useful in generating a long-term schedule such as those prepared for electric generation units in a large utility company. A systematic approach for turnaround maintenance is presented together with some specific examples of tasks to be performed in TA maintenance.

Appendix

- ***Objective function***

Maximize

$$\begin{aligned}&Y_{111}+Y_{122}+Y_{133}+Y_{144}+Y_{155}+Y_{166}+Y_{211}+Y_{222}+Y_{233}\\&+Y_{244}+Y_{311}+Y_{322}+Y_{333}+Y_{344}+Y_{355}+Y_{366}+Y_{377}+Y_{411}+Y_{422}\\&+Y_{433}+Y_{444}+Y_{555}+Y_{466}+Y_{511}+Y_{522}+Y_{533}+Y_{544}+Y_{555}+Y_{566}\\&+Y_{577}+Y_{611}+Y_{622}+Y_{466}+Y_{633}+Y_{644}+Y_{655}+Y_{666}+Y_{711}+Y_{722}+Y_{733}\\&+Y_{744}+Y_{755}+Y_{555}+Y_{766}+Y_{777}+Y_{522}+Y_{811}+Y_{822}+Y_{833}+Y_{844}+Y_{855}+Y_{866}\end{aligned}$$

Subject to

- Job-hour balance constraints

$$Y_{111}+Y_{122}+Y_{133}+Y_{144}+Y_{155}+Y_{166}=1$$
$$Y_{211}+Y_{222}+Y_{233}+Y_{244}=1$$
$$Y_{311}+Y_{322}+Y_{333}+Y_{344}+Y_{355}+Y_{366}+Y_{377}=1$$
$$Y_{411}+Y_{422}+Y_{433}+Y_{444}+Y_{455}+Y_{466}=1$$
$$Y_{511}+Y_{522}+Y_{533}+Y_{544}+Y_{555}+Y_{566}+Y_{577}=1$$
$$Y_{611}+Y_{622}+Y_{633}+Y_{644}+Y_{655}+Y_{666}=1$$

Skill-hour balance constraints

$$Y_{711}+Y_{722}+Y_{733}+Y_{744}=Y_{755}+Y_{766}+Y_{777}=1$$
$$Y_{111}+Y_{811}+Y_{822}+Y_{833}=Y_{844}+Y_{855}+Y_{866}=1$$
$$Y_{211}+Y_{611}\leq 2$$

$$Y_{122}+Y_{222}+Y_{311}+Y_{511}=Y_{611}+Y_{622}\leq 2$$
$$Y_{133}+Y_{233}+Y_{322}+Y_{411}=Y_{522}+Y_{622}+Y_{633}+Y_{811}\leq 2$$
$$Y_{144}+Y_{244}+Y_{333}+Y_{422}=Y_{533}+Y_{633}+Y_{644}+Y_{822}\leq 2$$
$$Y_{155}+Y_{344}+Y_{433}+Y_{544}=Y_{644}+Y_{655}+Y_{833}\leq 2$$
$$Y_{166}+Y_{355}+Y_{444}+Y_{555}=Y_{655}+Y_{666}+Y_{844}\leq 2$$
$$Y_{366}+Y_{455}+Y_{566}+Y_{666}+Y_{855}\leq 2$$
$$Y_{377}+Y_{466}+Y_{577}+Y_{866}\leq 2$$
$$Y_{311}+Y_{511}+Y_{711}\leq 2$$

$$Y_{111} + Y_{211} + Y_{322} + Y_{522} + Y_{722} + Y_{811} \leq 1$$
$$Y_{111} + Y_{122+}Y_{222+}Y_{333=}Y_{533+}Y_{733+}Y_{822} \leq 1$$
$$Y_{122+}Y_{133} + Y_{233} + Y_{344} = Y_{544} + Y_{744} + Y_{833} \leq 1$$
$$Y_{133} + Y_{144} + Y_{244} + Y_{355} = Y_{555} + Y_{755} + Y_{844} + \leq 1$$
$$Y_{144} + Y_{155} + Y_{366} + Y_{566} = Y_{766} + Y_{855} + \leq 1$$
$$Y_{155} + Y_{166} + Y_{377} + Y_{577} + Y_{777} + Y_{866} \leq 1$$
$$Y_{166} \leq 1$$

$$Y_{411} + Y_{811} \leq 1$$
$$Y_{411} + Y_{422} + Y_{711} + Y_{822} + \leq 1$$
$$Y_{211} + Y_{422} + Y_{433} + Y_{611} + Y_{722} + Y_{833} \leq 1$$
$$Y_{211} + Y_{222} + Y_{433} + Y_{444} + Y_{622} + Y_{733} + Y_{844} \leq 1$$
$$Y_{211} + Y_{222} + Y_{233} + Y_{444} + Y_{455} + Y_{633} + Y_{744} + Y_{855} \leq 1$$
$$Y_{222} + Y_{233} + Y_{244} + Y_{455} + Y_{466} + Y_{644} + Y_{755} + Y_{866} \leq 1$$
$$Y_{233} + Y_{244} + Y_{466} + Y_{655} + Y_{766} \leq 1$$
$$Y_{244} + Y_{666} + Y_{777} \leq 1,$$

Y_{ill} are zero-one variables

Exercises

1. What are the objectives of maintenance planning and scheduling?
2. State the elements of effective planning procedure.
3. What are the tools and methods that a planner could utilize for effective planning? (Mention 4).
4. What are the elements a reliable schedule must take into consideration?
5. What are the elements of sound scheduling?
6. List three techniques you can use for optimizing the maintenance scheduling process.
7. List the three benefits of the critical path method (CPM).
8. List the drawbacks of CPM.

9. Given the elements of a major maintenance job in the following table:

No.	Job	Normal time	Crash time	Normal cost	Crash cost	Job that precede the job	No. of crafts needed for the job
1	A	5	4	10	15	0	2
2	B	7	5	15	20	0	3
3	C	8	5	20	25	B	3
4	D	5	4	15	25	C, A	2
5	E	10	6	40	50	0	5
6	F	7	5	15	25	E, D	3
7	G	12	8	30	40	B	5
8	H	5	4	10	20	F, G	4

(a) Develop the network diagram for the job.
(b) Schedule the jobs so that the manpower is even, as much as possible.
(c) Put the schedule on a Gantt chart.

10. Extend the integer programming model by incorporating special equipment availability. Demonstrate the extended model by an example.

References

1. Bedworth DD, Bailey JE (1987) Integrated production control systems, 2nd edn. Wiley, Hoboken
2. Campbell JD (1995) Uptime: strategies for excellence in maintenance management. Productivity Press, Portland
3. Duffuaa SO, Al-Sultan KS. A mathematical model for effective maintenance planning and scheduling proceedings of the 2nd scientific symposium on maintenance planning and operations. Riyadh, pp 184–193
4. Duffuaa KS, Al-Sultan (1997) Mathematical programming approaches for the management of maintenance planning and scheduling. J Qual Maintenance Eng (JQME) 3(3):163–176
5. Duffuaa SO, Al-Sultan KS. A stochastic programming model for scheduling maintenance personnel. A working paper, Department of Systems Engineering, College of Computer Sciences and Engineering, KFUPM, Dhahran 31261 Saudi Arabia
6. Escudero LF (1984) On maintenance scheduling of production units. Eur J Oper Res 9 (3):164–274
7. Hartman EH (1987) Maintenance productivity: Why It Is So Low and How To Improve It. In: Hartman E (ed) Maintenance management, Industrial Engineering and Management Press
8. International Labour Office (1992) Introduction to work study, 4th edn. George Kanawaty (ed)
9. Montgomery DC (1976) Forcasting and time-series analysis. McGraw Hill, New York
10. Moder JJ, Phillips CC, Davis EW (1983) Project management with CPM, PERT, and precedence diagramminh, 3rd edn. Van Nostand reinhold Company, New York

11. Roberts SM, Escudero LF (1983) Minimum problem-size formulation for the scheduling of plant maintenance personnel. J Optim Theory Appl (JOTA) 39(3):345–362
12. Taha H (1992) Operations research: an introduction, 5th Edn. Maxwell Macmillam
13. Worrall BM, Mert B (1979) Application of dynamic scheduling rules in maintenance planning and scheduling. Proceedings of the 5th international conference on production research, (ICPR), pp 260–264

Chapter 8
Maintenance Quality Control

8.1 Introduction

The development of a sound quality control system for maintenance is essential for ensuring high-quality repair, accurate standards, maximum availability, and equipment life cycle and efficient equipment production rates. Quality control as an integrated system has been practiced with more intensity in production and manufacturing operations than that in maintenance. Although the role of maintenance in the long-term profitability in an organization has been realized, the issues relating to the quality of maintenance output have not been adequately formulated. Possible reasons include the following:

1. The output of the maintenance department is difficult to define and measure.
2. Customer focus is lacking in maintenance as compared to production.
3. A large portion of maintenance is non-repetitive.
4. Work conditions vary more in maintenance work than in production.
5. Traditionally, maintenance has been regarded as a necessary evil and at best a secondary system driven by production. This viewpoint has led to the assignment of a low priority to improvement of maintenance activities.

The quality of maintenance output has a direct link to product quality and the ability for a company to meet delivery schedules. In general terms, equipment that is not well maintained or that is maintained with poor workmanship fails periodically, or experiences speed losses, or is reduced of precision and hence tends to produce defects. More often than not, such equipment drives manufacturing processes out of control. A process that is out of control or with poor capability produces defective products which amount to lower profitability and greater customer dissatisfaction.

A clear organization of the quality control function and the specification of its role (responsibilities) in the maintenance system should be emphasized by the organization's top management. The responsibilities include development of testing

S.O. Duffuaa and A. Raouf, *Planning and Control of Maintenance Systems*,
DOI 10.1007/978-3-319-19803-3_8

and inspection procedures, documentation, follow-up, deficiency analysis, and help in identifying training needs from the analysis of quality reports.

Maintenance managers and engineers need to be aware of the importance of controlling the quality of maintenance output. The establishment of maintenance testing and inspection standards and acceptable quality levels should be developed for all maintenance work. Documentation of maintenance procedures and inspection reports can provide tremendous opportunities for maintenance quality improvement. These opportunities can be realized by continuous improvement of the procedures, and the identification of training needs to enhance craft technical skills.

Maintenance activities are not repetitive, and large observations for such activities cannot be collected for statistical analysis. For such activities, process control techniques provide valuable tools for improving maintenance processes.

Organizations should strive to tie their maintenance activities to the quality of their products and services. Also, they should create a focus on their internal customers. This will provide them with direction and goals for improving their maintenance processes.

This chapter presents the elements of maintenance quality control. Section 8.2 presents the responsibilities of quality control departments within maintenance; Sect. 8.3 describes inspection and testing programs; Sect. 8.4 presents the use of statistical process control in maintenance; Sect. 8.5 provides a program for improving maintenance jobs; Sect. 8.6 presents quality circles in maintenance; Sect. 8.7 outlines the link between maintenance and the quality of products and services provided by an organization; and Sect. 8.8 summarizes and concludes this chapter.

8.2 Responsibilities of Quality Control (QC)

Meeting maintenance quality and reliability objectives is the responsibility of all maintenance personnel. The combined effort and dedication of the QC personnel, maintenance supervisors, foremen, and technicians are essential to guarantee high-quality maintenance and equipment reliability. The quality control (QC) unit or function is responsible for ensuring the quality of maintenance output and the improvement of the maintenance process. Specifically, QC responsibilities include the following:

1. Performing inspections of maintenance actions, procedure, equipment, and facilities.
2. Maintaining and upgrading maintenance documents, procedures, and standards.
3. Ensuring that all units are aware and proficient in maintenance procedures and standards.
4. Maintaining a high level of expertise by keeping up to date with the publications concerning maintenance procedures and records.

5. Providing input in the training of maintenance personnel.
6. Performing deficiency analysis and process improvement studies using various statistical process control tools.
7. Ensuring that all the technical and management procedures are adhered to by crafts when performing actual maintenance.
8. Reviewing the job time standards to evaluate their adequacy.
9. Reviewing material and spare parts quality and availability to ensure availability and quality.
10. Performing maintenance audits to access the current maintenance situation and prescribe remedies for deficient areas.
11. Establishing certification and authorization of personnel performing highly specialized critical tasks.
12. Developing procedures for new equipment inspections and test the equipment prior to acceptance from vendors.

In summary, quality control in maintenance is responsible for ensuring the quality objectives for resources, procedures, and standards used in the maintenance process are met. In addition, it performs inspection of maintenance jobs and tests of equipment prior to acceptance or operation.

It is essential to have the QC personnel as independent as possible, and they must not be an extension of the workforce. Also, they should not perform production inspections, as such inspections can be assigned to production inspectors or workshop supervisors. Personnel comprising the quality control unit must be highly qualified technicians or engineers with extensive training in areas such as productivity improvement, statistical process control, process improvement, techniques for planning and scheduling, and work measurements.

In large organizations such as airline companies, air forces, army units, and railroad companies, it is necessary to have a quality control division within the maintenance department. This division will report to the maintenance manager. In medium-size organization, a small unit will do the job; however, in small-size organizations, one or two inspectors attached to the manager's office or the planning unit can perform the function of quality control.

The quality control department should develop a vision and a philosophy on quality management. An example of such a philosophy similar to the one adopted by the US Army is stated as follows:

> Product quality is largely dependent upon the skill and attitude of the repairman, the effectiveness of the supervisor, and the degree of compliance with standards and procedural instructions. Quality cannot be inspected into an item: it must be built-in by the individual repairman during shop processing. The quality of work performed is the responsibility of the repairman and supervisory personnel. Quality is directly influenced by the skill, attitude and motivation of maintenance personnel performing the work. It is inherent in the work process that every repairman checks his work to determine that quality specifications are met.

8.3 Inspection and Check Programs

The quality control department is responsible for performing and maintaining inspection records. The QC department should classify the different types of inspections it performs. Typically, QC performs several types of inspection. These inspections include the following:

1. Acceptance inspection. This type of inspection is performed to ensure that equipment condition is in accordance with standards. It is usually performed for new equipment.
2. Quality verification inspection. This type of inspection is performed following an inspection or a repair task to check whether the inspection or repair is done according to specification.
3. Document or file inspection. This is performed to review the standard and to evaluate its applicability.
4. Activity inspection. This is performed to investigate whether maintenance units are adhering to procedures and standards.

The above four types of inspections are performed by the QC personnel. There are other types of inspection that are performed by the maintenance production supervisors. These include inspections performed by the supervisors to ensure that material or workmanship meet prescribed standards.

The key figure in most maintenance quality control programs is the technical inspector. This is the individual assigned the responsibility of precluding the use of any defective workmanship and overcoming organizational deficiencies or reducing the unnecessary replacement of still serviceable components. Thus, the diagnostic ability of the inspector must be adequate for a successful QC program. The inspector is decisions are critical and may lead to catastrophe, especially in the airline industry.

8.4 Statistical Process Control (SPC) in Maintenance

SPC is the use of statistically based techniques to evaluate a process or its output to achieve or maintain a state of control. This definition is broad enough to include all statistically based methods ranging from data collection and histograms to sophisticated techniques such as design of experiments [2–4, 9]. Although there is no single list for these statistical methods (quality tools), there is a general agreement on the following seven tools, all of which require data collection as a first step:

1. Checklist
2. Histogram,
3. Cause and effect (fishbone) diagram,
4. Pareto chart (also known as ABC analysis),

5. Control charts,
6. Scatter diagram, and
7. Failure mode and effect analysis (FMEA).

8.4.1 Data Collection

The scientific approach for problem solving and the utilization of the above tools require the availability of the right data. Great care must be exercised in collecting the right data with the right method. To avoid repeating the process of data collection, delaying the analysis and the process improvement, the following guidelines are useful:

1. Plan the whole process of data collection at the beginning.
2. Clarify the purpose of data collection.
3. Specify clearly the data needed.
4. Use correct sampling techniques.
5. Design the needed checklists in advance; data gathering should be a continuous process and must be embedded in the available information system. Examples of the data needed in the case of maintenance management and engineering are downtime of equipment, labor productivity, maintenance costs, material and inventory costs, equipment failure and repairs, job completion times, and backlog.

8.4.2 Checklist

A checklist is a set of instructions, prepared in a simple manner, to aid in data collection, so data can be compiled, easily used, and analyzed automatically. There are many forms of checklists. Some of them are simple steps to perform maintenance inspection tasks (Fig. 8.1), while others could be a part of a lengthy auditing scheme. Checklist can be used in maintenance for the following:

1. Collecting data for developing histograms.
2. Performing maintenance tasks.
3. Preparing before and cleaning after maintenance jobs.
4. Reviewing spare parts.
5. Planning maintenance jobs.
6. Inspecting an equipment.
7. Auditing a maintenance department.
8. Checking defective causes.
9. Machine diagnosis.
10. Collecting data for work sampling.

Examples of checklists are given in Figs. 8.1, 8.2, 8.3, and 8.4.

		Check the column that indicates the condition of the unit									
Number of major components	Item	Good Condition	Requires Cleaning	Requires Adjustment	Requires Lubrication	Examine Vibration	Examine Heat	Loose	Requires Overhaul	Requires Replacement	See Additional Comments
1	Electric Motor										
	1.1 Bearing										
	1.2 Base and Bolts										
	1.3 Temperature										
	1.4 Vibration										
	1.5 Noise										
2	Coupling										
	2.1 Alignment										
	2.2 Lubrication										
3	Generator										
	3.1 All Electric Motors										
	3.2 Armature										
	3.3 Brushes										
	3.4 Rotor										

Fig. 8.1 Preventive maintenance inspection checklist for a typical motor–generator set

8.4.3 Histogram

A histogram is a graphical representation of frequency of occurrence versus data points or a class that represents a set of data points. It is a graphical picture of the frequency distribution. The histogram helps the viewer visualize the data distribution, shape, and dispersion. One major use of histograms is to identify the distributions of data. It can be used to estimate the following:

1. The maintenance load,
2. Spare parts supplier reliability,
3. Equipment time to failure distribution,
4. Distribution of repair times,
5. Distribution of backlog, and
6. Shift in downtime distribution.

In summary, it can be used for identifying distribution of important activities. Any statistical software such Statgraphics [14] or Statistical Analysis Software (SAS) [6] provides capabilities for computing frequency distributions and plotting histograms. Following is an example of the use of histograms in maintenance.

Assume that in a machinery shop, a grinder that is used for finishing machine parts was found to break repeatedly. The effect of this recurring breakdown caused

Work order # :_____________________Date :___________________

No.	Major process operation	Tasks	Time in	Time out	Time of ** non-regular breaks
1.	Review work scope	1.1 Review work			
2.	Set-up, check run-out	2.1 Check			
		2.2 Tail Stock			
		2.3 Steady Rest			
		2.4 Check run out			
3.	Premachining	3.1 Premachining			
		3.2 Steam cleaning			
4.*	Plasma or HVOF spray	4.1 Preheating in oven			
		4.2 Fixing shimms			
		4.3 Gritting			
		4.4 Preheating using gun			
		4.5 Spraying			
		4.6 Cooling down			
5.*	Brush spray	5.1 Preparation			
		5.2 Platting			
		5.3 Final Machining			
6.*	Metal Spray	6.1 Preheating			
		6.2 Spraying			
		6.3 Final Machining			
7.	Grinding	7.1 Grind			
8.	Probe Tracking	8.1 Grind			
		8.2 Burnishing			
		8.3 Measuring and taking photos			

* = Only one of 4,5, or 6 is applicable

** = A non-regular break refers to an assignment or consultation on a different job, change in shift, or over night break. Regular breaks include coffee, tea, bath room, and consulting with supervisor and technical staff on the same job.

Fig. 8.2 A checklist to standardize a shaft repair process

Physician :____________________			Date :__________
Patient :____________________			Age :__________
Circle for Required test	Test	Result	Normal Values
1.	Urea N		7 - 50 Mg/dL
2.	Glucose fasting		60 -110 Mg/dL
3.	Uric Acid		2 - 7 Mg/dL
4.	Creatinine		0.05 - 1.5 Mg/dL
5.	Amylase		45 - 200 Mg/dL
6.	Sodium		136 - 144 mEq/L
7.	Potassium		4 - 4.5 mEq/L
8.	Chloride		95 - 105 mEq/L
9.	CO_2		23 - 30 mM/L
10.	Calcium		8.5 - 10.4 Mg/dL
11.	Cholesterol		128 - 200 Mg/dL
12.	Triglyceride		33 - 210 Mg/dL
13.	Laclate dehydrogenase (LDH)		65 - 115 Mg/dL
14.	High density lipoprotein (HDL)		
15.	Acid phosphates		Less 1.0 I.U
16.	Serum Protein		6 - 8 GM/dL
17.	Serum albumin		3.5 - 5.0 GM/dL
18.	Serum globulin		23 - 35 GM/dL

Mg/dL = milli-gram per deci litter, m eq/L = milli equivalent per liter

m M/L = milli-Mole per Liter, I.U = International Unit

GM/dL = grams per deci Liter N = Nitrogen

CO_2 = Carbon Dioxide

Fig. 8.3 A checklist for blood diagnosis

Work Sampling Check Sheet

Checker: ________________ Object of Check ____________ Date: __________

Method: ________________ Weather: ________

Item	Checks	Total	%
Processing	## ## //----------	300	60%
Planning	## ## //----------	125	25%
Transport	## // --------------	50	10%
Break-down	## ## /	20	4%
Others	## //	5	1%
Total		500	100%

Fig. 8.4 Work sampling check sheet

serious problem, such as production delays, late deliveries to customers, lower productivity, and high defect rates. The daily reports from the shop floor showed the hours worked and the amount produced, and noted the grinder downtime for not meeting production plans. These daily reports showed the time of the grinder breakdown and the time it was restarted after it had been repaired. The data for downtime is given in Table 8.1.

A frequency distribution of the data in Table 8.1 is given in Table 8.2.

A histogram for the downtimes using the classes given in Table 8.2 is shown in Fig. 8.5.

Table 8.1 Grinder downtime in hours for the last three months (rounded to the nearest half hour)

1, ½, ¾, 2, ½, ½, 1, 1, ¾, 3, 1, ¾
1, ¾, 2, 8, 3, ¾, ½, 8, 4, 7, 10, ¾
2, ½, 1, ½, 30, 1, 2, 1, 50, 6, 3, 3
5, 4, 6, 5, 72, 2, 4, 2, 3

Table 8.2 Frequency distribution for the downtime data in Table 8.1

Breakdown duration	Frequency	Relative frequency
$0 < t \leq 1$	20	44.44
$1 < t \leq 2$	6	13.33
$2 < t \leq 3$	5	11.11
$3 < t \leq 4$	3	6.67
$4 < t \leq 5$	2	4.45
$5 < t \leq 6$	2	4.45
$6 < t \leq 7$	1	2.22
$t \geq 8$	6	13.33
Total	45	100.00

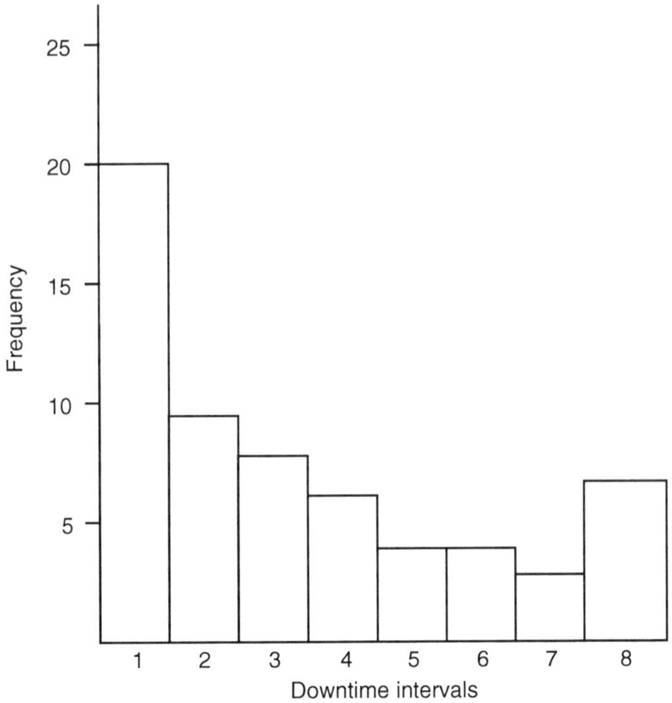

Fig. 8.5 A histogram for the data in Table 8.2

The histogram shows the following:

1. It took over 8 h between stopping and starting production on six occasions.
2. Most of the stoppage was not very long.
3. The mode of the downtime is ½ hour.

The next step in maintenance quality improvement is to identify the causes of breakdown. A valuable tool for that is the cause and effect diagram.

8.4.4 Cause and Effect (Fishbone) Diagram (CED)

A cause and effect diagram is usually used as a tool to show reasons for substandard maintenance performance. CED has been utilized extensively in production processes control, and it is also known as the fishbone diagram. It is useful in sorting out causes and organizing mutual relationships. The effect is usually taken to be the quality characteristic needs to be improved and the causes are the influencing factors. It can be used in maintenance management and engineering to identify the causes of the following:

1. Low craft productivity.
2. Excessive downtime.
3. Recurring breakdown.
4. Repeat jobs.
5. Excessive absentees.
6. Backlog.
7. Excessive errors in data recording.

The steps of constructing the CED is stated and demonstrated on the grinder case given in Sect. 4.1.

Step 1: Decide the quality characteristic and the effect needed to be studied. This is usually a phenomenon (effect) we need to improve and control. In our case, it is the grinder downtime;

Step 2: Write the effect on the right side. Draw a broad arrow from the left side to the right side (Fig. 8.6);

Step 3: Write the main factors which may be causing the grinder downtime, directing a branch arrow to the main arrow (Fig. 8.7). Group the major possible causes into categories such as materials, equipment, methods of work, and measuring methods. Each category will form a branch as in Fig. (8.7);

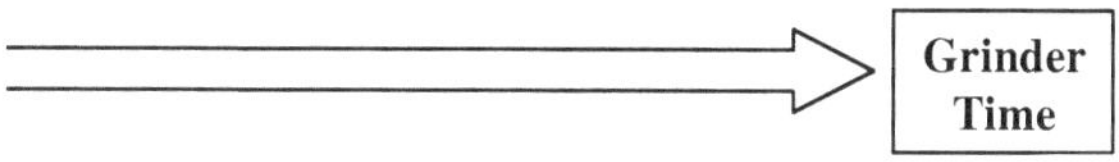

Fig. 8.6 Main *Arrow* for the Cause and Effect Diagram

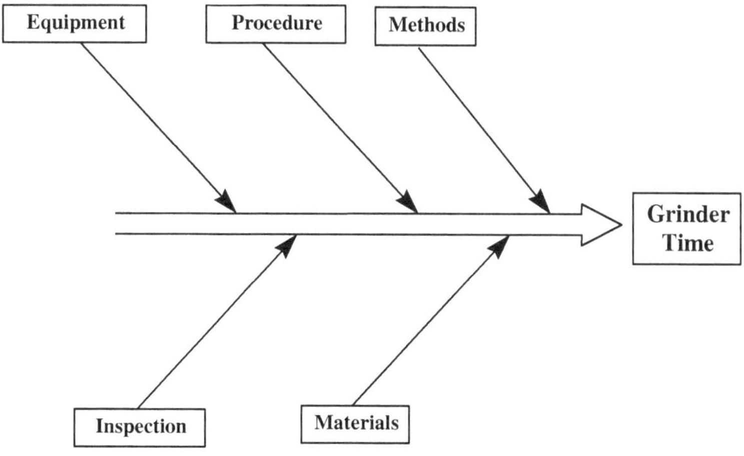

Fig. 8.7 Major Categories of Causes for the Diagram

Step 4: Write on each of the side branches the detailed factors which may be regarded as the causes. These will be like twigs and on these write more detailed factors and so on (Fig. 8.8); and

Step 5: Check to make sure that all causes are included in the diagram and the relationships are properly illustrated.

It was found after examining the record that the most frequent types of breakdowns are the following:

1. Uneven rotation,
2. Does not rotate,
3. Rise in coolant temperature,
4. Polishing powder does not come out,
5. Coolant leakage, and
6. Others.

When focusing on reasons for long downtime, it was found the reason for that include the following:

1. lack of inspection procedures to detect failure,
2. long delays in communicating failure,
3. the nonexistence of a standard procedure for planning jobs,
4. long repair times, and
5. excessive breakdown.

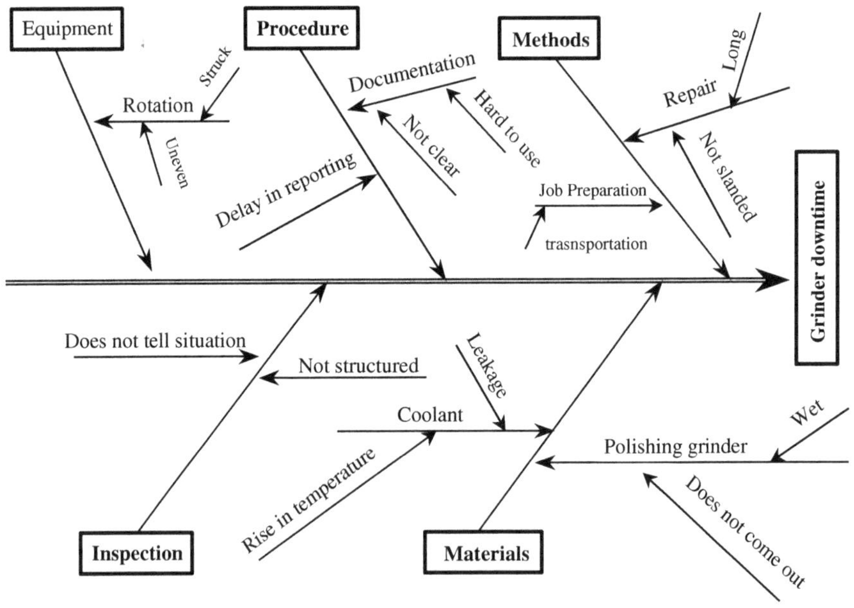

Fig. 8.8 Complete cause and effect diagram for grinder downtime

One option for reducing downtime is to develop effective procedures for detecting failure and reporting it. Another approach is to eliminate breakdowns. The second approach seems to be a better one. The question is which breakdowns should be to eliminated? It is natural to eliminate the breakdowns which are happening most frequently and therefore reduces the time used for reporting and repairing. An alternative approach is to eliminate the breakdown causing the most downtime, if the data are available. A common technique that is used to prioritize options is the Pareto chart also known as ABC analysis. The next section explains this technique and uses it to identify the most frequent breakdown.

8.4.5 Pareto Chart (ABC Analysis)

A Pareto chart is simply a frequency distribution of attribute data arranged by order of frequency. Its purpose is to identify the vital few from the many trivial. It also helps in setting priorities as to which course of action is most beneficial. For example, in maintenance engineering, there are many factors that could be improved. These are craft productivity, equipment uptime, equipment quality rate, spare parts availability, etc. It is impossible to have the resources to improve all of these factors. A Pareto chart can be of help in identifying the important factors for improvement and based on that priorities can be established as which factor to improve.

A Pareto chart indicates which factor to improve first in order to eliminate defects and reap the largest possible improvement. It classifies the factors into three classes. Class A usually contains around 20 % of the factors (causes) that are causing 75–80 % of the problem. Class B contains factors causing 15–20 % of the problems. The rest of the factors (which are many) are in class C.

The construction of a Pareto chart is as follows:

Step 1: Divide the data in items or classes you will use in the graph;
Step 2: Establish the time horizon for the chart;
Step 3: Determine the frequency of each item or class. Sort these according to frequency in a descending order; and
Step 4: Plot item or class versus frequency starting with the largest frequency and continuing in descending order. In the same graph, you can plot cumulative frequency versus item or classes.

For more on the construction of the chart see [5, 9].

The chart may be used for the same example of the grinder breakdown. Assume the goal is to identify the most common breakdowns. Examine the work orders of the 45 breakdowns given in Table 8.1. Frequency is shown in the graph given in Fig. 8.9.

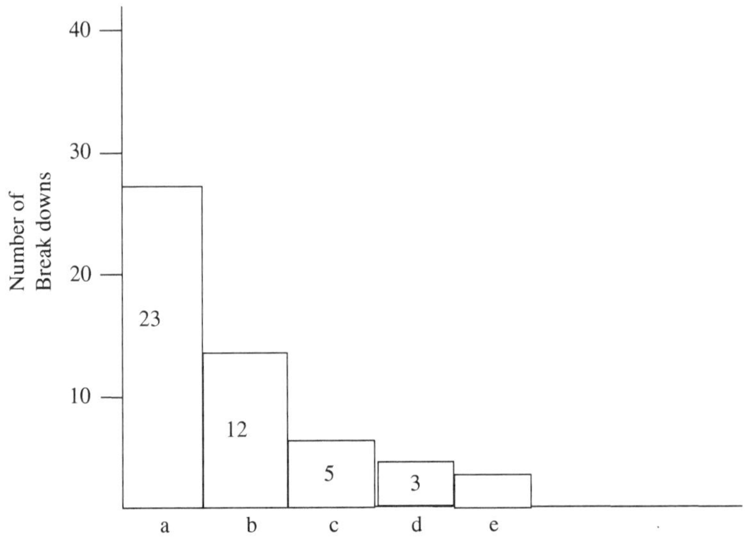

Fig. 8.9 A Pareto chart for grinder breakdown. **a** Polishing powder does not come out. **b** Rise in coolant temperature. **c** Uneven rotation. **d** Does not rotate. **e** Others

In other applications of maintenance, Pareto charts can be utilized to identify:

1. factors impairing productivity,
2. crafts causing major backlog,
3. spare parts causing most delays,
4. the most costly spare parts, and
5. breakdown causing the most downtime.

8.4.6 Control Charts

Control charts are important in statistical process control and are used extensively in quality control. Control charts can be applied to improve maintenance activities, and they differ from the others techniques described earlier. The previous techniques are static, and control charts are dynamic, allowing observation of a process over time. The construction of the control charts is explained in Chaps. 4 and 6.

Control charts can be used to monitor the quality of the following items:

1. Monthly backlog,
2. Downtime of major equipment,
3. Equipment availability,
4. Equipment quality rate, and
5. Number of breakdown.

In addition, control charts can be used to access adequacy of maintenance standards and monitor tool wear. In case of tool wear, a moving range chart has been developed for determining both the region of tool failure and the time of expected failure. Control charts have been used to establish a threshold for condition maintenance. In this application, a control chart is established based on sensor measurements and the control limits capture the machine's normal vibration and can be used to track faults.

8.4.7 Scatter Diagram

The scatter diagram is a graphical relationship between two variables, and it shows the correlation between them. It is usually used to study the relationship between causes and effects. It therefore complements the cause and effect diagram mentioned earlier. In general, it may be applied to carry out the following analyses:

- Trend analysis and
- Correlation or pattern analysis.

Particularly in maintenance, it can be utilized to find the following:

1. Correlation between preventive maintenance and quality rate.
2. Correlation between level of training and backlog.
3. Correlation between level of training and repeat jobs.
4. Correlation between vibration level and quality rate.
5. Correlation between preventive maintenance and downtime.
6. Downtime trend.
7. Trend of maintenance cost.
8. Trend of crafts productivity.
9. Backlog trend.
10. Equipment availability trend.

8.4.8 Failure Mode and Effect Analysis (FMEA)

Failure mode and effect analysis is a technique used to quantify and rank critical failures in product or process design. It involves identifying all functional and cosmetic characteristics. Then, for each characteristic, FMEA identifies a list of potential failures and their impact on the overall product performance. Also, the likelihood and the severity of the failure (problem) are estimated.

For a given system or a product, it is broken down into basic components or parts. Then, for each component or part, the reader may ask how it could fail, the probability of failure, and the effect on the product or system function. In an example of electronic assembly, the parts include each resistor and capacitor analog device. Then, the mode of failures for each part will be identified and its effect on product functions. The objective is to remove the threat to product integrity or minimize it.

This technique has been applied successfully in product design in the automotive industry and in the selection of critical factors for design of experiments in quality engineering. It has a great potential for use and application in maintenance, especially in accessing the effect of failure modes on functional failures when designing a reliability centered maintenance program (RCM). It can be applied in selecting a design modification for a system in operation.

8.5 Maintenance Job Quality Control

Maintenance work differs from production work because it is mostly non-repetitive and has more variability. For non-repetitive, sometimes work enough data cannot be collected in order to use SPC tools effectively. In these cases, it becomes essential to control the process of maintenance through the control of its inputs.

A process is a sequence of steps (operations) that transforms a set of inputs to an output, and it also has a feedback mechanism. Figure 8.10 depicts the definition of a process.

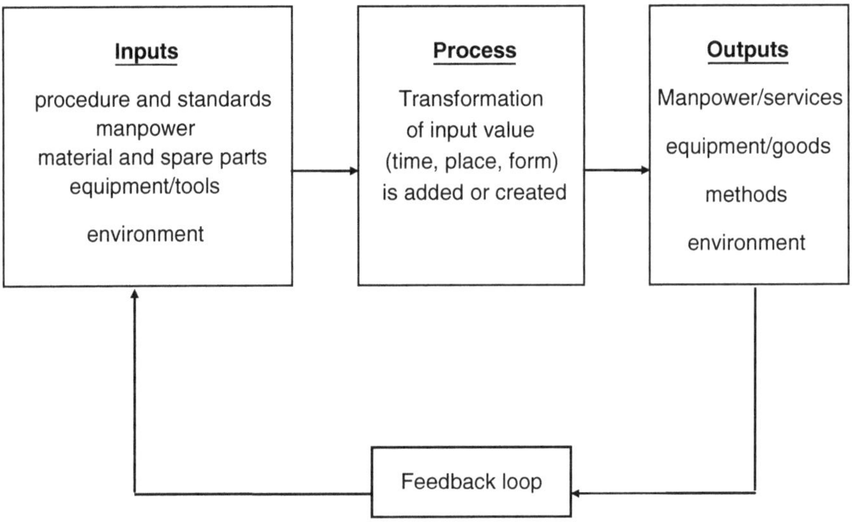

Fig. 8.10 Definition of a process with feedback loop

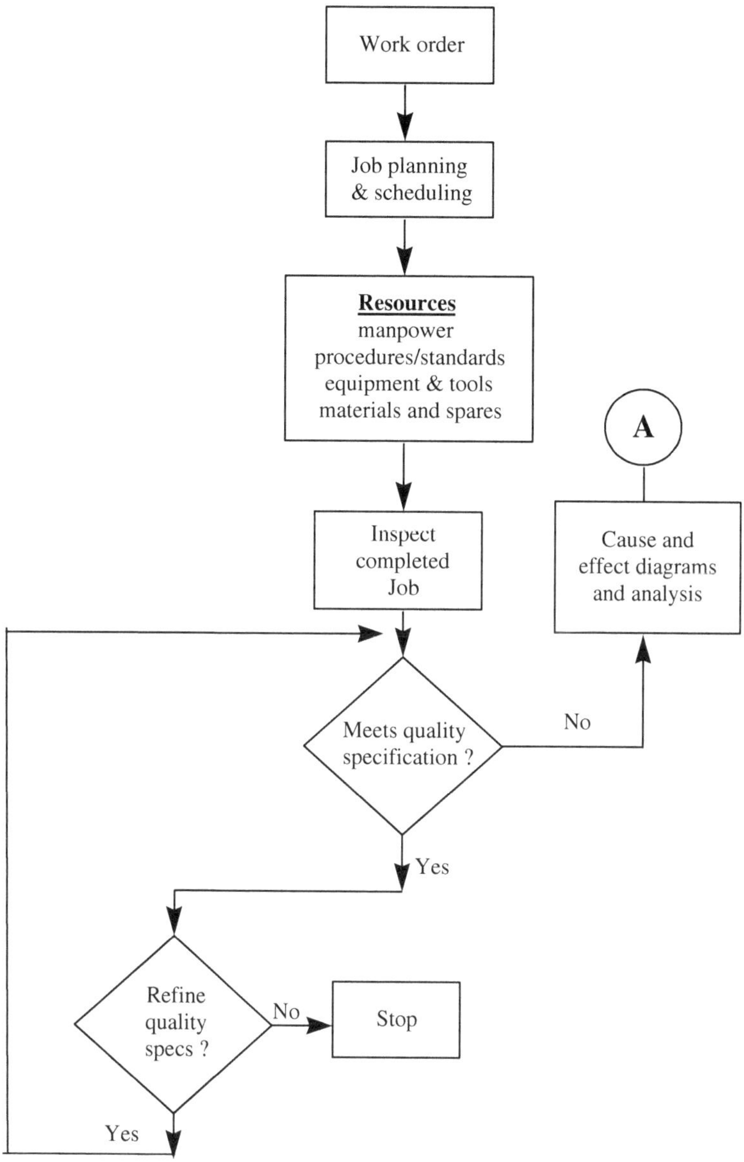

Fig. 8.11 Steps for controlling and improving maintenance job quality

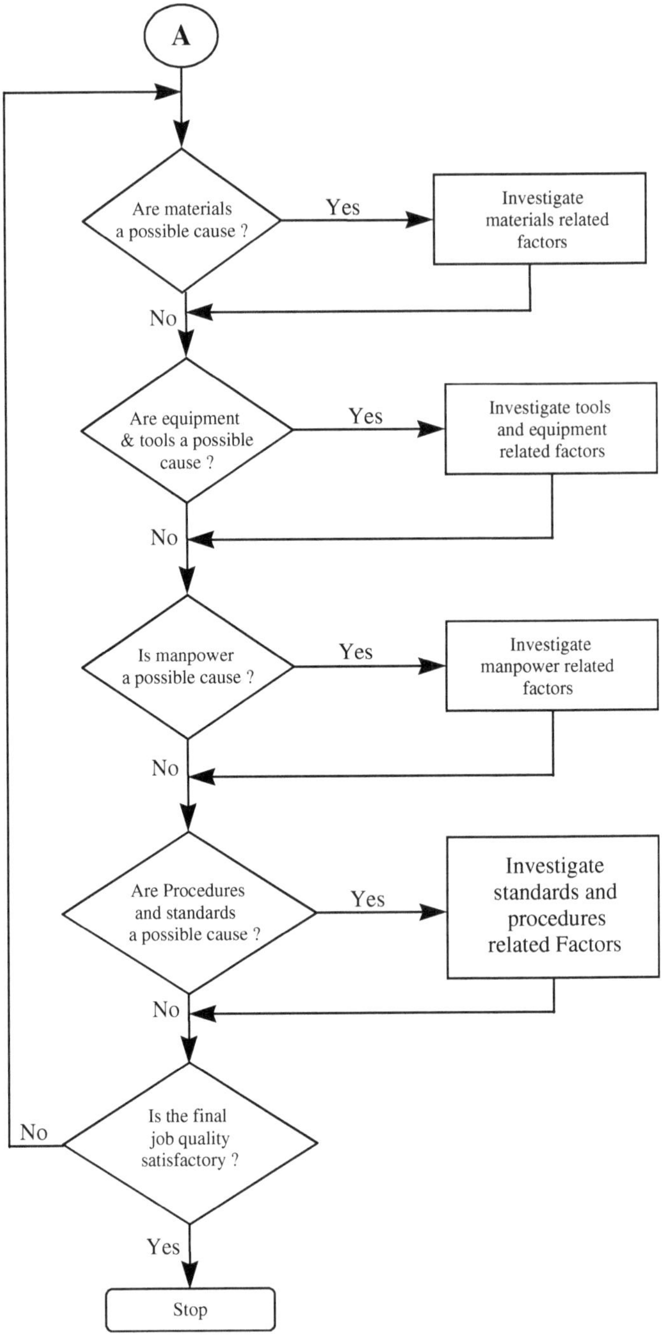

Fig. 8.11 (continued)

The major inputs to the maintenance process in order for a job to be performed are as follows:

1. Maintenance procedures and standards,
2. Manpower,
3. Material and spare parts, and
4. Equipment and tools.

The above four inputs are critical for maintenance job quality. The key element for maintenance job quality is to develop job quality standards for critical non-repetitive jobs. Then, if a job does not meet the standard, a cause and effect diagram is used to investigate the root causes for the substandard jobs. Figure 8.11 shows the steps for controlling and improving maintenance job quality. The steps in the figure are the first to discover the cause of low quality (in terms of meeting specifications and time standard) and must be coupled with further investigations using cause and effect diagrams for other inputs to the maintenance jobs. Figure 8.11 provides example for this analysis in case the procedures and standards are suspected to be the most likely case for substandard job quality. In the next subsection, we focus on the major factors that affect maintenance job quality and provide a procedure for accessing their impact (Fig. 8.12).

8.5.1 Factors Related to Procedures and Standards

Procedures and standards are prescribed to control the work and ensure uniformity and quality. In order for the standards to ensure quality, they must be precise, measurable, and reflect customers (here, the operation is considered as an internal customer) requirements. A procedure should be clear, logical, and well documented in order to be implemented. The following are the major factors that affect procedure and standards effectiveness.

1. Procedure quality (its capability to meet customer requirements),
2. Procedure and standard documentation,
3. Adequacy of standards for job environment, and
4. Mechanism for procedures and standard improvement.

The quality of a procedure is evaluated based on its ability to achieve its objectives. The subfactors used in evaluating any procedure include objective clarity, logical structure, clarity, simplicity, easiness to follow, comprehensiveness, specification of responsibility, computerization, documentation, and mechanism for continuous improvement.

Standards are evaluated by testing their adequacy.

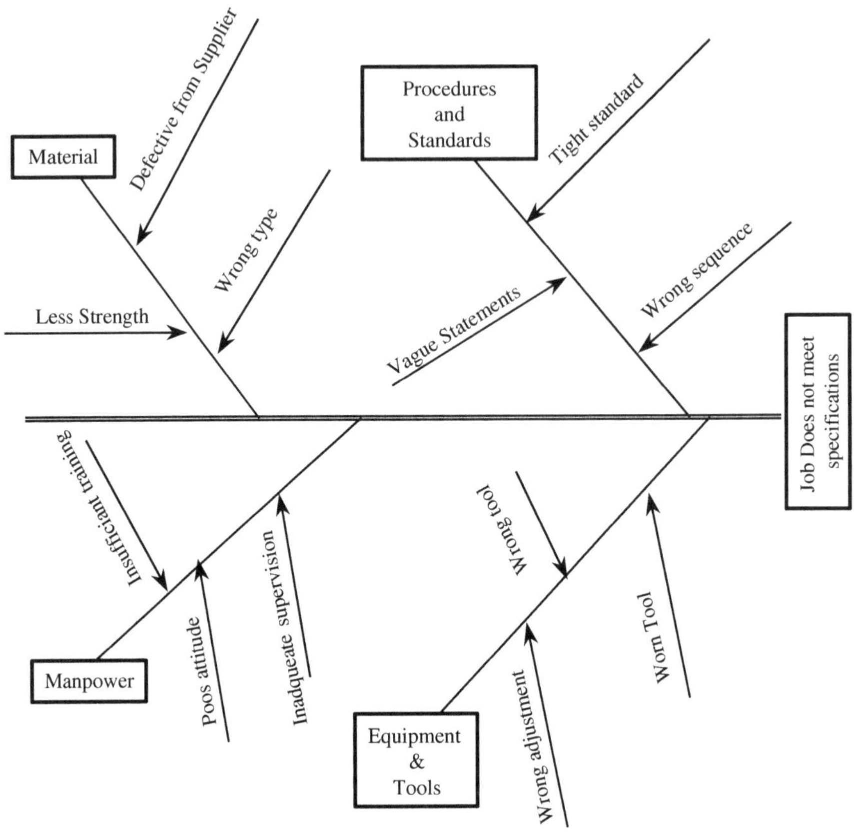

Fig. 8.12 Cause and effect diagram to investigate jobs not confirming to standards

8.5.2 Factors Related to Manpower

The role of qualified manpower is essential for high-quality maintenance. Still, the qualified technician plays a key role in maintenance. The following are the most important factors to monitor in order to improve maintenance job quality: workforce size, skill, training, motivation, attitude, work environment, and background.

8.5.3 Factors Related to Material

The availability of material in the right quality and quality at the right time contributes to maintenance job quality. The factors that affect the availability and the quality of material include correct standards and specifications, material control policies, budget, purchasing policies and procedures, and material handling and deployment. These factors constitute the basis for cause and effect analysis to investigate the impact of materials and spares on maintenance job quality.

8.5.4 Factors Related to Tools and Equipment

The availability of equipment and tools for performing production maintenance can be a limiting factor in some circumstances. For example, the accuracy of calibration and precision instruments could have a big impact on the quality of maintenance jobs. The factors that affect the availability of the right equipment and tools include budget, operational readiness, training on use, compatibility, and number of equipment on hand. These factors could constitute a starting list for a brainstorming session on the impact of the availability of equipment and tools on maintenance job quality.

8.6 Quality Circles

Quality circles are small groups which voluntarily perform quality control activities within their workshops. Moreover, these groups should be a continuing organization, within company-wide quality control activities (CWQC). Quality circles were started first in Japan in 1962 and have been practiced in many countries outside Japan including the USA and many European countries. The basic concept behind quality circles activities within the framework of a company-wide quality control effort is as follows:

1. To contribute to the improvement and development of the enterprise.
2. To respect humanity and improve quality of life.
3. To enhance human capabilities to its fullest capacity and draw out industrial potential.

Those objectives are met through implementing the following:

1. Performing quality control activities within the circle's workshop. The circle activities encompass all the departments in the company including production, maintenance, and service activities.
2. Strive for self-development. This implies that the quality circle members must study and attempt to develop in them the ability to solve problems.
3. Participate in the effort of mutual development and the dialogue among quality circles by presenting successful case studies in quality circles conferences.

Quality circles can play a positive role in improving the quality of maintenance activities. The following case demonstrates the role of quality circles in maintenance. This case reports on a successful case study performed by a quality circle at Kimitsu Works, a steel plant at Kimitsu, Japan, to improve equipment maintenance devices so that they could be used more widely and effectively to diagnose machine failures. As machines and other equipment become bigger and more complex, it is impossible to maintain equipment effectively with a simple "feel" for the job and trial and error method. Moreover, the need to reduce failures and equipment problems require that equipment diagnosis to be done in scientific ways to yield quantifiable values. The company has 53 machine checks (MC) that were purchased to help in effective diagnosis; however, these checkers were not fully utilized. A study was done to find their utilization rate and it was found to be 30 %. It was believed that if the utilization rate is improved that would help in more effective machine diagnosis and thus reduce equipment failures and problems. The purpose of this case was to enhance utilization, and a target of 80 % utilization was set.

The survey conducted also inquired about the reasons the workers are not utilizing the machine checker. A Pareto chart was constructed to identify the major causes (reasons) for not using the checker. Only three causes were responsible for the low utilization of the checker. The reasons were given as follows: (1) instructions for use of the checkers are inconsistent, (2) workers did not know how to use the checkers, and (3) evaluation criteria were hard to use.

The quality circle group brainstormed on the three major causes and developed a plan to improve utilization and divided the responsibilities for implementation. The plan included the development of a new easier performance evaluation device and establishment of a nucleus group to study the theory and practice of using machine checkers in order to become leaders in training sessions for the use of machine checkers. For a detailed description of the case refer to [8].

8.7 Maintenance Link to Quality

Maintenance can contribute significantly in enhancing and maintaining quality products; for example, the capability of a machine tool that is in top condition will produce more than 99 % of the parts within the accepted tolerance. Figure 8.13a illustrates the distribution of product quality characteristics. After the machine has been in service for sometime and wear has taken place in some of the machine parts, there will be more chatter and vibration. The distribution of the quality characteristic will have more variation and more parts will be produced that are out of specification. In addition, more parts will have particular quality characteristic far away from the target value for those characteristic. The new distribution is shown at the bottom graph in Fig. 8.13b. In general terms, a process that is out of control produces defective products and therefore increases production cost that amounts to lower profitability which endangers the survival of the organization.

Predictive and condition-based maintenance employ a closed-loop maintenance strategy in which information from equipment is obtained and utilized in making planned maintenance decisions. The maintenance decision is usually based on the use of a threshold which, when reached, means that maintenance is to be carried

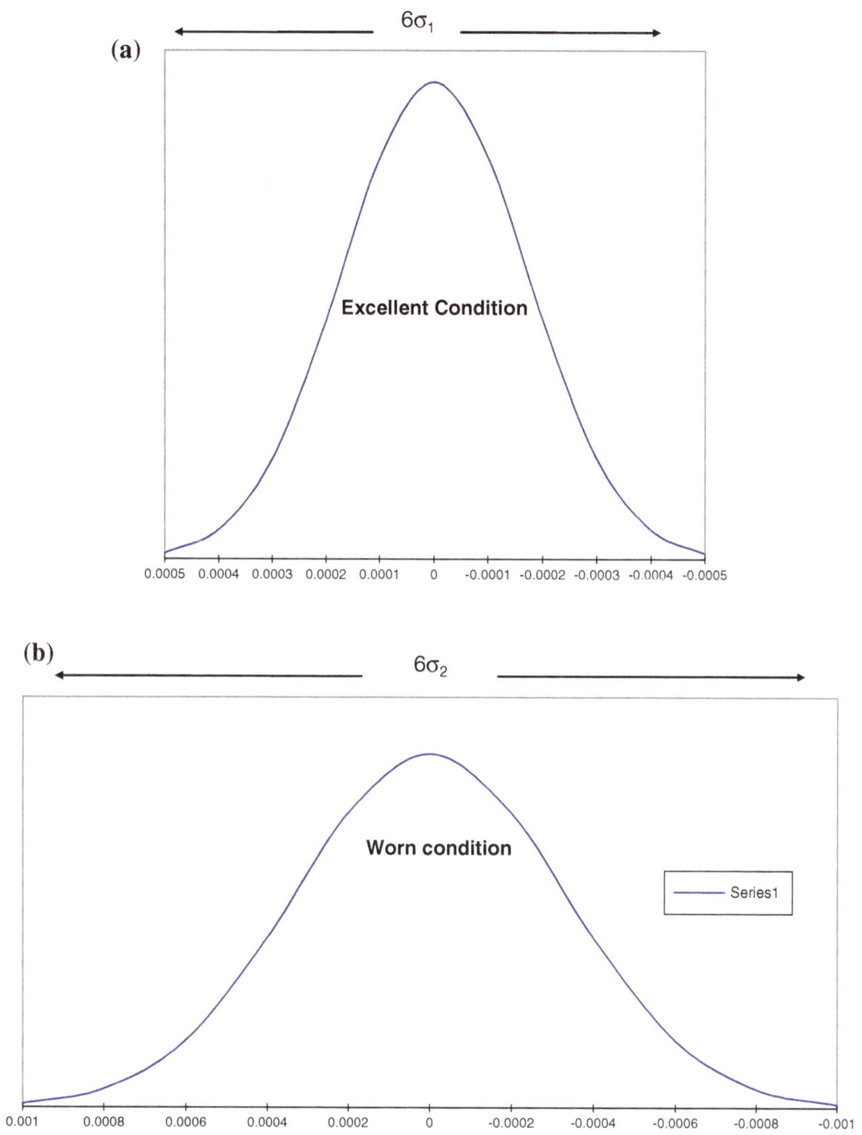

Fig. 8.13 Distribution of a product quality characteristic for new and worn machine

out. Such a strategy will ensure high product quality, especially if these thresholds are chosen so that equipment does not deteriorate to the extent at which defective, or near defective, products are generated.

The above arguments demonstrate the strong link between equipment maintenance and product quality. The total product maintenance (TPM) philosophy has also identified the link. To achieve overall equipment effectiveness, TPM works

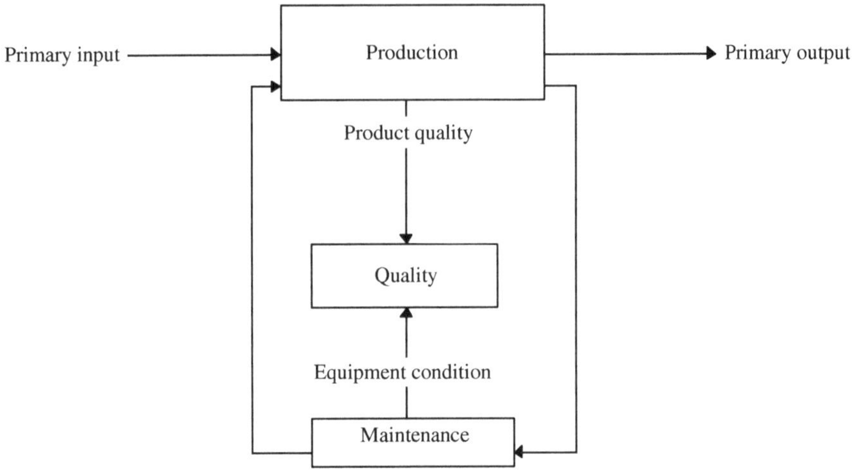

Fig. 8.14 Relationships among production, quality, and maintenance

toward minimizing equipment losses by minimizing equipment failure, setup and adjustment, idling and minor stoppage, reduced speed, process defects, and reduced yield. All these losses are related directly or indirectly to quality. For example, faulty equipment tends to produce defects prior to reaching complete failure. Setup and adjustments are usually made to find the optimal parameter settings. It is implied that adjustment is done because the equipment is not operating at optimal conditions, and therefore, defective products are expected to be produced. Also, at the setup and adjustment stage, defects are produced until the optimal adjustment is found. Idling, minor stoppage, and reduced speed often produce defects, since product parameters are a function of speed. For example, the surface finish in machining is a function of speed. For more on TPM and the relation between overall equipment effectiveness and quality see Chap. 14.

Maintenance is a system that operates in parallel with production. The primary output of production is the desired products with certain level of quality that is defined by the customer. As the production process continues, it generates a secondary output which is the demand for maintenance and an input to the maintenance process. The output of maintenance is equipment in serviceable conditions. Well-maintained equipment increases the production capacity and represents a secondary input to the production. Therefore, maintenance affects production by increasing production capacity and controlling the quality and quantity of output. The relationship and the link ages among production, quality, and maintenance have been studied in detail in [1], and a framework for modeling the link was outlined using the imperfect maintenance and Taguchi methods. Figure 8.14 depicts the relationships among production, quality, and maintenance.

8.8 Summary

In this chapter, different approaches for maintenance quality control and process improvement are presented. The essence of this chapter is to stress the benefit of using statistical process control tools for improving all aspects of the maintenance process. Examples demonstrating the use of these tools in maintenance have been provided. The link between maintenance and quality has been outlined. Integration and the utilization of these tools and the availability of accurate data are important for realizing improvement. The maintenance management information system should include the needed data for maintenance quality control. This may require the merger of the maintenance management information system with the quality information system. For more on the subject of maintenance quality control and improvement see [7, 10–15].

Exercises

1. What are the reasons for not formalizing the quality of maintenance output?
2. What are the output of the maintenance function and specify how to measure it?
3. List the responsibilities of maintenance quality control.
4. What are the types of inspections performed by maintenance quality control?
5. List five possible uses for checklists in maintenance.
6. List six possible uses for the histogram in maintenance.
7. Visit an organization near your area and collect some data on a maintenance activity and use it to demonstrate the use of control charts in maintenance.
8. What is the role FMEA can play in improving the quality of maintenance?
9. Visit an organization near you and identify a substandard maintenance job due to materials and use the cause and effect diagram to obtain the possible root causes for this problem.
10. Provide two examples to demonstrate the link between quality and maintenance.

References

1. Ben Daya M, Duffuaa SO (1995) Maintenance and quality: the missing link. J Qual Maintenance Eng 1(1):20–26
2. Ben Daya M, Raouf A (1996) A revised failure mode and effects analysis model. Int J Qual Reliab Manage 13(1):43–47
3. Duffuaa SO, Ben Daya M (1995) Improving maintenance quality using SPC tools. J Qual Maintenance Eng 1(2):25–33
4. Everly JR (1983) Maintenance quality control: a critical appraisal: a technical report. US Army War College, Carlisle Barracks, pp 17–13
5. Ishikawa K (1976) Guide to quality control, vol 107. Asian Productivity Organization Tokyo, Japan

6. International Business Machine (1984) Statistical analysis software, New York
7. Leonard SC (1993) Improve maintenance reliability. Hydrocarbon Process 72(8):C162
8. Masuda T, Taniguchi H (1976) Working for better equipment maintenance. In: Ishikawa K (ed) A case study appeared in quality control circles at work
9. Montgomery DC (1991) Statistical quality control, 2nd edn. Wiley, New York
10. Nakajima S (1988) Introduction to TPM. Productivity Press Inc., Norwalk
11. Niebel BW (1994) Engineering maintenance management, 2nd edn. Marcel Dekker Inc, New York
12. Osuri RH, ChaHerjee S, Chandrashekhara K (1991) On line monitoring of tool wear in end milling using acoustic environment. Int J Prod Res 29(7):1339–1353
13. Peace GS (1995) Taguchi methods a hands-on approach to quality engineering. Addison Wesley Publishing Company, New York
14. Statistical Graphics Corporation (1989) Statgraphics. STSC Inc, USA
15. Yoshinori I (1988–1989) Reducing equipment trouble, the association for overseas scholarship (AOTS), vol 110. Winter, USA, pp 26–31

Chapter 9
Maintenance Training

9.1 Introduction

Maintenance activity is considered to be a function inclusive of actions that are necessary for keeping equipment and facilities performing at desired levels. Among all the factors necessary, maintenance training programs are the most important. It is imperative that maintenance personnel acquire the requisite technology and skills which will enable them to exert their full potential.

Maintenance workers are traditionally craft oriented and have more freedom of action than production workers. They require extensive training and experience to be fully qualified. This coupled with the fact that maintenance is dynamic, in that, new deficiencies in equipment are continually arising, while old problems are in the process of being corrected. Also, that new equipment based on the latest technology is being inducted by industry, makes it very imperative that maintenance worker's training be planned, carried out and its effectiveness evaluated.

Efforts are made to assign the right workers to different functions so that each worker is fully conversant with his/her functions. In fact, workers available are never entirely qualified. A lack of formal training is one of the main reasons for that. Workers need to be trained, but the question is, how much and in what areas?

This chapter introduces skill and skill levels, type of training, a framework for indicating an appropriate training program, and means of evaluating the effectiveness of such training programs. Section 9.2 covers skill levels, and Sect. 9.3 covers maintenance training activities. Section 9.4 summarizes this chapter.

S.O. Duffuaa and A. Raouf, *Planning and Control of Maintenance Systems*,
DOI 10.1007/978-3-319-19803-3_9

9.2 Skill

Skill may be defined as the ability to perform one's job. Skills needed by maintenance workers are quite different than the skills needed by production workers. Figure 9.1 shows a conceptual model of human performance.

Production work is routine and, as such, requires a lesser amount of information to be processed than maintenance work. Maintenance work presents different levels of information processing and decision making. It is considered that performance of maintenance workers improve through a combination of motivation and training.

9.2.1 Skill Levels

It is necessary to identify existing skill levels of workers before a training program is initiated. There are four levels of skill:

- Level 1: Person lacks theoretical knowledge and practical ability
- Level 2: Person is conversant with theory but lacks practical training
- Level 3: Person possesses practical experience but lacks theoretical concepts
- Level 4: Person is adequately conversant with both theoretical aspects and has practical competency.

A true training program should be designed to meet the above-cited needs. To make training effective, due care must be exercised in scheduling the appropriate contents of the training and scheduling it at the right time.

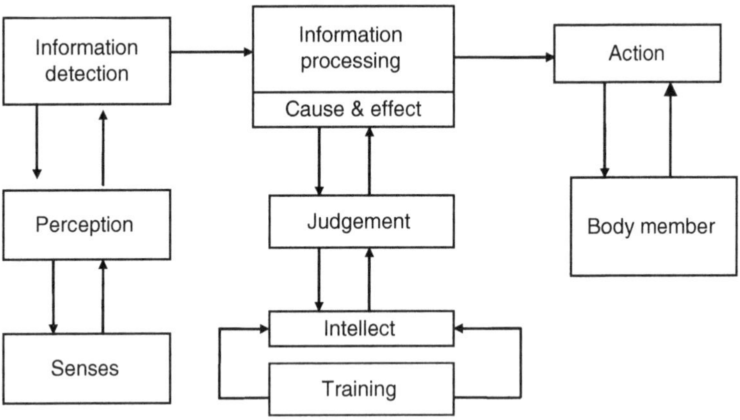

Fig. 9.1 A conceptual model of human performance

9.3 Maintenance Training Activities

Maintenance training activities can be considered in the form of a cycle as shown in Fig. 9.2.

9.3.1 Training Policy

Most of the companies give their employees some form of training. The existing training program must be reviewed to see its effects on raising skills of maintenance workers. The focus should be on developing a training program that improves the existing conditions. Issues such as developing multi-skilled workers, imparting specialist skills through a combination of lecture, on-the-job training, have to be clearly stated in the basic policy. To develop equipment-competent maintenance and to nurture human resources that will meet long-term requirements are two important objectives that must be stated under the training program goals. Priorities have to be set very clearly for developing equipment-competent people, administration-competent people, and the development of an ability-development system.

The policy should state clearly the vision, the long-term objectives of the training program and the organization's priorities in training. An effective mechanism to achieve the objectives of the program must be developed.

9.3.2 Evaluating Current Status

Equipment is becoming more sophisticated and automated. This increases the need for assured quality of output, low energy consumption, and safety of operation. A well-designed training system is very essential so that the objectives stated are met.

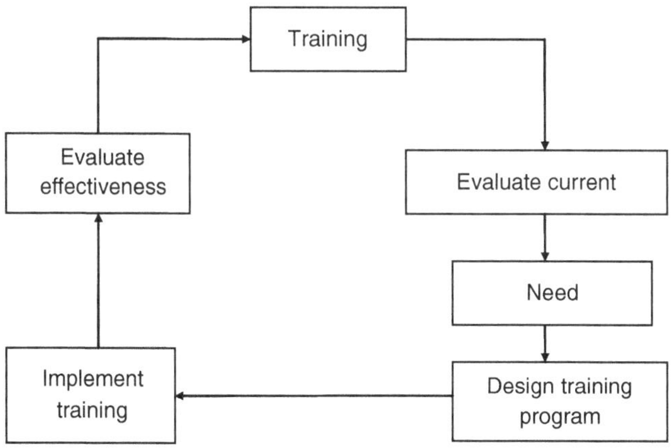

Fig. 9.2 Maintenance training activities

Usually, maintenance professionals should be able to perform the following activities. This, by no means, is an exhaustive list of such activities.

1. To assess whether the equipment is operating normally
2. To diagnose the causes of abnormalities and restore normal operations
3. To improve equipment reliability and minimize abnormalities and failures
4. To minimize the related costs.
5. Perform the job with the required level of quality and safety.

Keeping in view the functions to be performed, an inventory of skills possessed by the maintenance personnel should be taken. This will help in developing the needed training program. For taking the inventory of skills, it is essential that a list of typical tasks that the person is performing on his/her machine and future potential tasks to be done be prepared. A specimen craft skill inventory form is shown in Fig. 9.3.

On this form, typical tasks to be performed on the equipment, along with future potential tasks, are listed. Skill levels required for each task are listed on it as well.

CRAFT SKILL INVENTORY FORM

Equipment name : ____________________ DATE: _____

LOCATION: __________________

Name of assessor: ____

Work classification	Skill level required	Craft #1 (name)		Craft #2 (name)		Craft #3 (name)		Craft #4 (name)		Total need
		Skill level	Addition	Skill level	Addition	Skill level	Addition	Skill level	Addition	
Equipment Operation	4	3	1	3	1	3	1	3	1	4
Tracing Abnormalities of equipment mal-function	4	3	1	3	1	3	1	3	1	4
Equipment diagnostics	4	3	1	3	1	3	1	3	1	4
Safety	4	2	2	3	1	4	0	2	2	5
Background knowledge	3	2	1	3	0	3	0	2	1	2
Total			6		4		3		6	9

Fig. 9.3 Craft skill inventory form

On the same form, worker's name and current skill levels are entered. The current skill level is usually provided by the supervisor and may be determined by interviewing the worker. The completed form can be analyzed to estimate training requirements in a quantified manner. This form enables categorizing task-related training needs and additional training needs of workers.

9.3.3 *Need Analysis*

It is essential that a worker should possess necessary skills for performing a given task but is not sufficient to perform the tasks adequately. Factors such as motivation and availability of necessary tools are known to affect worker performance.

A need analysis is carried out to identify worker performance deficiencies, the causes of these deficiencies, and determine the appropriate solutions.

This process, normally, is based on the following five steps:

- Identifying thc desired performance,
- Identifying the deviations between expected and actual performance,
- Identifying root causes of the deficiency (ies),
- Identifying appropriate solutions, and
- Selecting and implementing the appropriate solutions.

9.3.3.1 Desired Performance

The desired performance of workers is known and predetermined. It usually involves but not limited to the following:

- Detecting equipment abnormalities,
- Correcting the abnormalities, and
- Maintaining the equipment.

9.3.3.2 Deviations Between Actual and Expected Performance

The difference between the expected and actual performance can be judged by the rework that a maintenance worker has to do, quality of equipment, output, etc.

9.3.3.3 Root Cause(s)

Usually, the reason of the performance deficiency falls in one of these categories.

Fig. 9.4 Interaction of factors

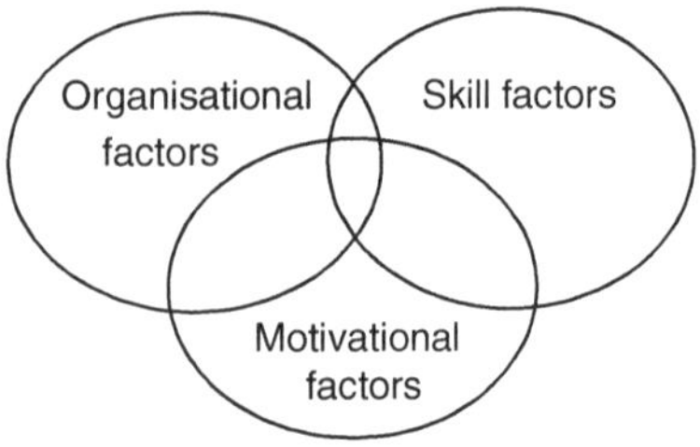

- **Knowledge and skill factor**

Worker may not be able to do his/her job because he/she does not know how because of a lack of required knowledge and skill.

- **Organizational factors**

Worker may know how to perform his jobs but does not have the required tools, equipment, and references.

- **Motivational factors**

Worker may know the job, has everything he/she needs, but may not be motivated to perform the task up to the required standards

The possible interactions of all these factors are shown in Fig. 9.4. While determining the root causes, their interaction must be kept in mind.

9.3.3.4 Solution

Identifications of root cause(s) of the performance deficiencies lead to the solutions. In addition to the training, the following may also be kept in mind:

- job standard may be examined,
- tools and equipment to be provided, and
- necessary incentives be provided to workers.

9.3.3.5 Selection

After the appropriate solutions have been identified, these should be implemented. Training is commonly neglected and postponed. To avoid this, a proper training program should be formalized.

9.3.4 Designing Training Program

Currently, acceptance of robots, numerical control machines, and flexible manufacturing systems by the industry is increasing every day. Maintenance of this

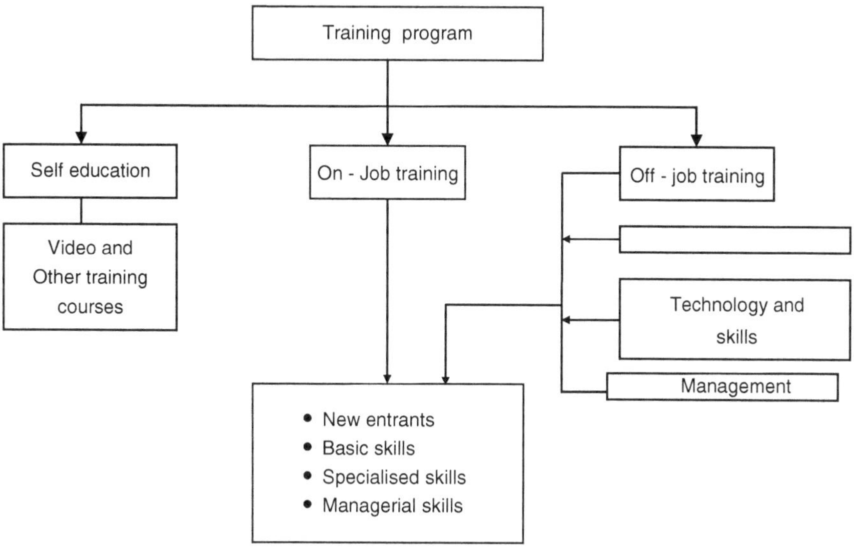

Fig. 9.5 An outline of a training program

sophisticated equipment poses a challenge to the maintenance group and also increases demands of multi-skilled personnel. Engineers and technicians, in addition to maintenance workers, have also to be trained. It is desirable that a company develops equipment competency at every level. To develop such competency, a program of training must be constructed that progresses in steps from elementary through basic, intermediate, and advanced skills. An outline of a training program is shown in Fig. 9.5.

It is desirable that each company devises its own particular system to suit its equipment. Companies usually have neglected equipment maintenance training in the past. Due consideration must be given to improving maintenance skills from the basic level on up. Instructional methods adopted by the companies do influence the pace of learning and retention of training information by the trainees.

9.3.5 Implementation of Training Program

A training curriculum is to be developed. While developing such a curriculum, due consideration must be given to the equipment that the company has, skill levels required for maintaining it, and what specific items to teach and how much time should be spent in teaching these items. A typical maintenance training system may have the following courses coupled with on-the-job training as shown in Fig. 9.6.

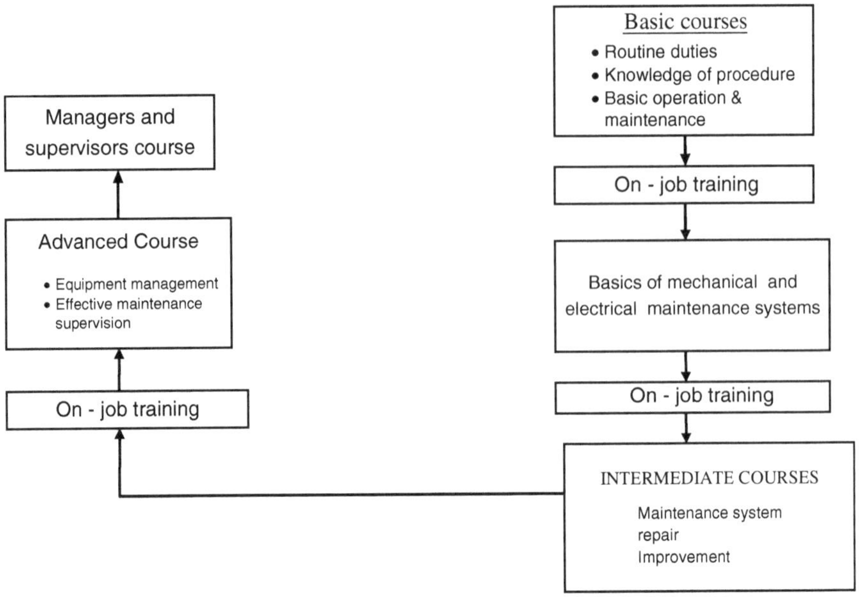

Fig. 9.6 A typical maintenance training system

9.3.6 Evaluating Effectiveness

The effectiveness of any training program should be evaluated periodically. During this evaluation, the progress made by individuals toward skill development is to be observed. Periodic reviews of the skills, training system, training processes, and the curricula are essential ingredients for continuous improvement. For this purpose, a craft skills inventory form can be used. This form provides the information regarding additional skills needed by each worker and also the total additional skills for all the workers with respect to maintaining particular equipment. The figures obtained before the onset of training program may be compared with the figures collected after completion of the program. Rates of progress may be examined to see whether the target has been achieved or not. This may necessitate looking at the entire program in case the desired target is not achieved. Similarly, rates of rework by each worker before training and after training will give some ideas regarding the effectiveness of the training program.

9.4 Summary

A brief outline of training has been provided. Each organization must develop its own training program. Skill levels have been described in very general terms, and these must be described by each company in very specific terms for each equipment. The maintenance training cycle has been described, and explanations of each step of this cycle are given. A framework for developing a training program and its implementation has been provided.

Means of evaluating effectiveness of training programs along with continuous improvement of effectiveness are briefly outlined. Efforts have been made to introduce training. For a detailed and traditional treatment of the subject, readers may refer to Refs. [1–5].

Exercises

1. Define the term skill. What are the possible skill levels in a typical maintenance department?
2. Flowchart the cycle of the maintenance training activities.
3. How would you carry out a need analysis to identify maintenance training needs?
4. What are the key factors that influence worker's job performance?
5. Explain how would go about designing a training program for maintenance.
6. How would you evaluate the effectiveness of maintenance training program?
7. Locate an organization in your area and evaluate their existing maintenance training program, and suggest improvements to it.

References

1. Baily RW (1992) Human performance engineering, 2nd edn. Prentice Hall
2. Hartmann EH (1992) Successfully installing TPM. TPM Press, Allison Park
3. Neible BW (1994) Engineering maintenance management. Marcel Deckker, New York
4. Pike RW (2003) Creative training techniques handbook, 3rd edn. HRD Press, Amherst
5. Weiss WH (1981) Performance based training plant engineering, May 1981

Chapter 10
Computerized Maintenance Management Systems

10.1 Introduction

The objective of maintenance organizations is to maximize uptime in the most cost-effective manner. To accomplish this objective, the following strategies must be clearly specified:

1. Effective maintenance strategies derived from equipment condition and history.
2. Efficient techniques for planning and scheduling of work orders and utilization of resources.
3. Monitoring of maintenance activities, data collection, and performance reporting to support continuous improvement.

The above three activities require information about equipment, crafts, work orders, jobs, job standards, production schedules, and the nature of operations in the organization. The amount of information that is collected, processed, and used for decision making is overwhelming. This necessitates a systematic approach for information management. Also, the complexity and the uncertainties involved in the process of maintenance and engineering and the amount of information handled in a typical maintenance system require computer support. Appropriate computer support provides the means for quick and timely response.

A Computerized maintenance management system (CMMS) is basically information systems adapted to serve maintenance [10]. CMMS aids in the process of data collection, recording, storing, updating, processing, communicating, and forecasting. It is essential for planning, scheduling, and controlling the maintenance activities. Through effective reporting, a CMMS can provide maintenance managers and engineers with the information needed for sound decision making to control and improve the maintenance process.

Most organizations today have some sort of computerized maintenance support and may have failed to reap the full benefits of a CMMS. The main reasons for that are the following:

S.O. Duffuaa and A. Raouf, *Planning and Control of Maintenance Systems*,
DOI 10.1007/978-3-319-19803-3_10

- In many cases, the system does not meet the maintenance requirements.
- The system is not user-friendly.
- Maintenance planners and engineers are not well trained to use the CMMS.
- CMMS reports are not used for maintenance improvement.

More analysis for reasons of failures in the use of CMMSs and possible remedies are given in [7]. In the process of developing an in-house system, or choosing one from the hundreds of available commercial packages, the above reasons for failure must be addressed carefully.

This chapter presents the elements of an effective CMMS and outlines a systematic procedure for their evaluation. Section 10.2 presents an overview of a CMMS. Section 10.3 describes the basic modules of a CMMS and presents flowcharts for the design and implementation of these modules. Evaluation of CMMS is given in Sect. 10.4. Section 10.5 outlines the evolution of CMMS for effective maintenance. Section 10.6 summarizes and concludes the chapter.

10.2 Overview of CMMS

The success of a CMMS can be measured by its ability to support the maintenance process. Two important elements are essential for an effective CMMS. The first one is its ability to support the main activities in the maintenance process, and the second is the ability of the software and hardware configuration in terms of their reliability, friendliness (ease) of use, quality of information, and timely processing.

A CMMS can be centralized for small organizations or completely decentralized and distributed systems for large ones. It can run on mainframe, microcomputer [1], workstations, and personal computers. Also, it can be a stand-alone system or a network in a client server environment. A typical PC-LAN (network) may have several terminals, remotely located with perhaps over fifteen users and several printers. The software can be menu-driven or window based. A typical CMMS is linked to inventory, payroll, purchasing, and accounting. Experience has shown that the system reliability, timeless, and friendliness enhance the utilization of CMMS and increase its benefits to the organization.

In terms of support to the maintenance process, CMMS usually includes the following functions:

- Equipment identification and bill of material
- Preventive maintenance
- Work order management
- Planning and scheduling

- Inventory control and purchasing
- Labor and job standard
- Equipment history
- Costs and budget
- Performance reports
- Quality reports

In the design process of a CMMS one or more of the above functions usually are grouped in one module. In this chapter, a CMMS design that supports the above function is presented. It consists of five models. These are as follows:

1. Equipment management
2. Work order control
3. Crafts management
4. Material supply and control
5. Performance reports.

10.3 CMMS Modules

The structure of a CMMS is given in Fig. 10.1. It consists of the above five modules that interact together to support the maintenance process. In addition to the basic structure given in Fig. 10.1, the CMMS has security and file creation systems. A similar structure to the one given in Fig. 10.1 has been developed and tested on IBM personal computers or compatibles [8]. The language used for software coding is dBase III (with limited modification, dBase III + can be used).

10.3.1 Equipment Management Module

This module provides information about equipment identification, location, installation date, status, technical information, equipment history and preventive maintenance schedule, special tools, and safety procedure. Additional information can be stored about lubrication program specifications, lubrication oil, and method of application, standard jobs, and repair history.

This module tracks equipment lubrication and preventive maintenance. It interacts with work order planning and controls in order to generate the needed work orders. Figure 10.2 shows the top page of an equipment file. The information available in this file can be used for making decisions regarding equipment replacement.

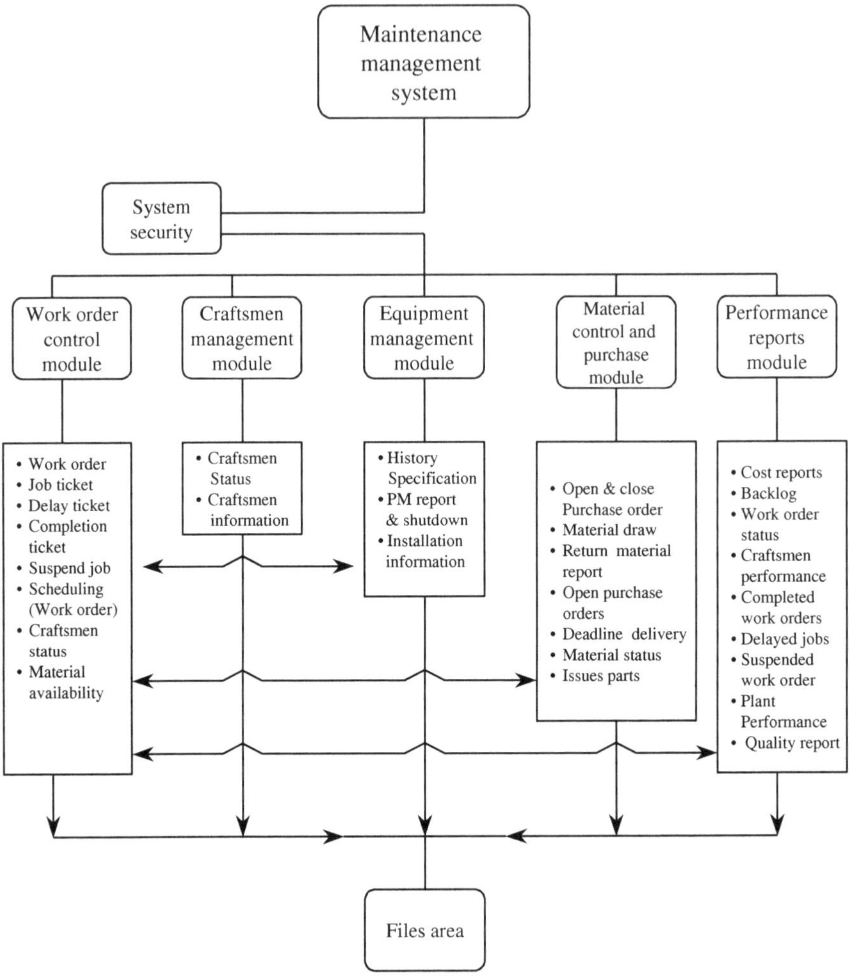

Fig. 10.1 Maintenance management information system structure

10.3.2 Work Order Control Module

The work order control module automates the process of work order generation. This module is the heart of the CMMS since it is responsible for the execution of the work order system, which is the heart of the maintenance control. The work order serves the following functions:

1. Documenting the process of work requests and authorizing the work to be performed.
2. Planning, monitoring, and controlling the actual work.

Equipment file

Equipment No. : 234567890122 Serial No. : 3JKJDF888875 Install. In : 11/11/95

Equipment location : SKDF 883 Equipment cost : 878698.44

Equipment name: Cutting machine

Specification : Voltage → 220/fixed location / area 3 sqm/
oil change after 720 hrs of usage/max. Work
hrs 18/day

Equipment code : 2

PM date : 12/12/96 Shutdown date : 11/11/93 Status (on/off) : on

Equipment code : 1 = util. 2 = produc. 3 = handel.sys.4 = facili. 5 = other

Equipment PM plan:

Tool needed in PM :

PM safety procedures:

Fig. 10.2 Equipment file

3. Helping in data collection about maintenance performance and costs.
4. Providing the needed information for feedback and continuous improvements.

The work order requires two types of information. The first type is used for planning and scheduling, and the second is for identification.

The details of work order design, function, and flow of information are given in Chap. 6. Figure 10.3 shows the flowchart for the work order control module. This module initializes the work order by feeding the necessary information regarding new jobs. The categories of information are listed below:

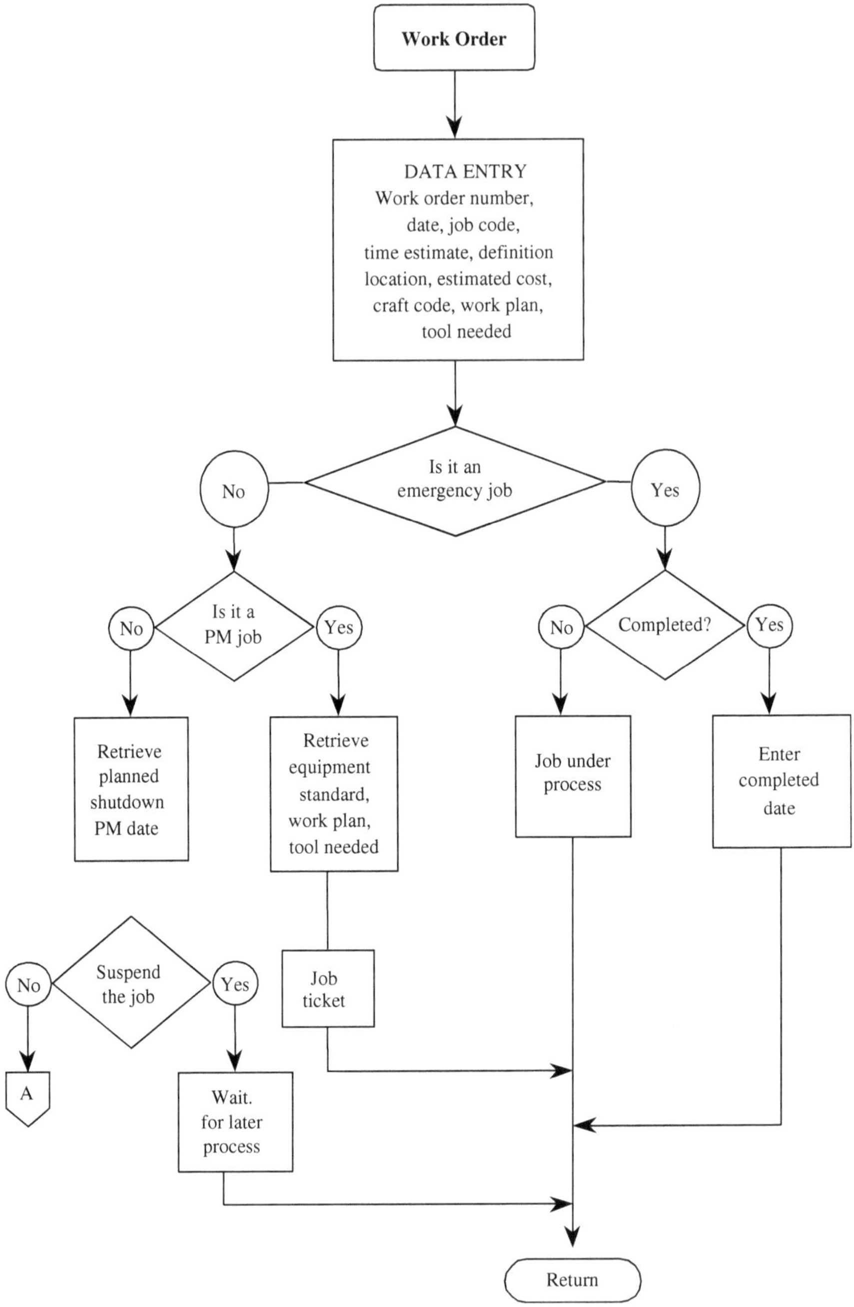

Fig. 10.3 Flowchart work order control

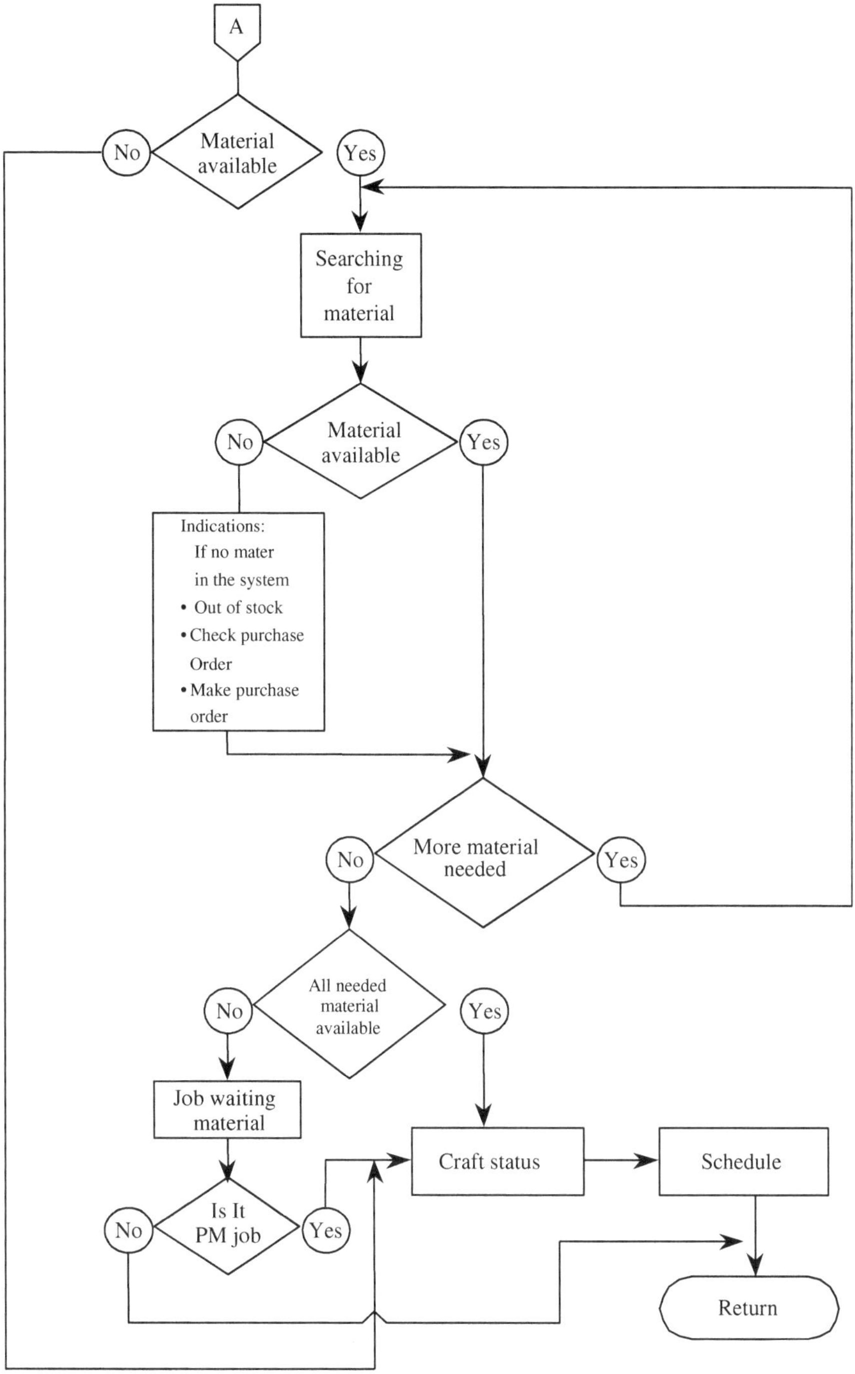

Fig. 10.3 (continued)

Table 10.1 Jobs and crafts codes

Job code		Craft code	
Code	Definition	Code	Definition
A	Emergency	1	Mechanic
B	Routine	2	Electrician
C	Preventive maintenance	3	Welder
D	Shutdown	4	Plumber
E	Other	5	Painter
		6	Carpenter

- Work order number
- Date
- Problem definition and tool needed
- Equipment number, work plan, and location
- Estimated man-hours and estimated costs
- Craft code
- Material

Table 10.1 illustrates job priority and craft code that indicate craft specialty. This module aids in the process of planning and scheduling. The first step for each work order is its planning. Then, after a group of work orders are planned, they are scheduled. A need exists to have a scheduling algorithm to be a part of the work order control module. In most systems, this feature is lacking. The description of such an algorithm is given in Chap. 6 on planning and scheduling. For each completed work order, a report is generated. This report provides an indication of the flow of work and may be used in a flowchart of job performance.

10.3.3 Crafts Management Module

This module keeps track of craft status in order to provide the planner/scheduler the information needed for scheduling work orders. The planner/scheduler, when executing the work order control module, needs to interface with this module to know manpower availability. The craft status report is given in Table 10.2 as an example for such information. Table 10.3 shows a personal file for a craft (Figs. 10.4 and 10.5).

10.3.4 Material Supply and Control Modules

Material requirement planning and inventory control are critical for overall performance of the maintenance system. A CMMS must provide an effective inventory and material supply.

Table 10.2 Weights for CMMS functions

Function	Weight out of 100 (W_i)
Work order management	25
Preventive maintenance management	15
Inventory management	15
Quality control management	15
Maintenance reports	10
Vendor evaluation	10
General considerations	10

Table 10.3 Characteristics of the work order management

No.	Characteristics of the system	Grade
1.	Corrective work orders can be produced	
2.	Preventive maintenance work orders can be produced	
3.	Tracks labor costs automatically	
4.	Tracks material costs automatically	
5.	The work order uses priority codes	
6.	The work order uses status codes	
7.	Sorts backlog by craft and priority	
8.	Produces a list of active work orders	
9.	Maintains an active equipment history	
10.	Allows for manual entry of work order cost estimates	
11.	Produces a list of work orders ready for scheduling	
12.	Provides net capacity calculations to compensate for work Interruptions	
13.	Allows for complex planning such as crafts and materials tools	
14.	Provides effective scheduling algorithms	
	Calculate total score $A_1 = W_1 \sum_{j=1}^{14} S_{1j}$	

In the process of planning a work order, the planner identifies the needed spares and material. Then, he checks the availability of spares and material in the needed amounts prior to opening the work order. The availability of spare parts and material is critical and essential for smooth planning, scheduling, and controlling of maintenance work. The details of methods for inventory and spares control are given in Chap. 7.

The system must have the capability of classifying spares according to their use and cost. An ABC analysis of all ledger maintenance stock and stores should be made. This type of analysis helps in identifying class A items which are the fifteen percent of all items that share the major inventory cost (70–80 % of the cost of all items). The class A items should effectively be controlled using a perpetual computer control system. Maximum and minimum levels for such items must be maintained and an economic order quantity should be used for restocking them.

CRAFTSMEN STATUS REPORT							
CRAFT CODE	NORMAL EMP.	HRS -HRS	UNAVAILABLE MED	VAC	STAFF ASSIGN	AVAILABLE STAFF/HR	SCHEDULED HRS.
ME	3	24	1	1	0	1/8	10
ELE	4	32	1	1	1	1/8	0
WELD	0	0	0	0	0	0/0	0
PLM	0	0	0	0	0	0/0	0
PAI	0	0	0	0	0	0/0	0
CAR	0	0	0	0	0	0/0	0

Fig. 10.4 Craftsmen status report

CRAFTMAN FILE

NAME : XXXXXXXXXXXXX | Date : 11/1195

CRAFT No. : 11111111111 | SALARY : xxxxxxx

JOB TITLE : Industrial Eng. | CRAFT CODE : X

ADDRESS/REMARK : ______________________

CRAFT STATUS : 1

STATUS → 1 = Assignment, 2 = Vacation, 3 = Medical, 4 = Onduty

Fig. 10.5 Craftsman personal file

Such supply modules can provide information about critical items and stock checks when needed. Typically, such modules are linked to purchasing in order to initiate purchase requisitions. The basic functions of the module given in Figs. 10.6 and 10.7 are the following:

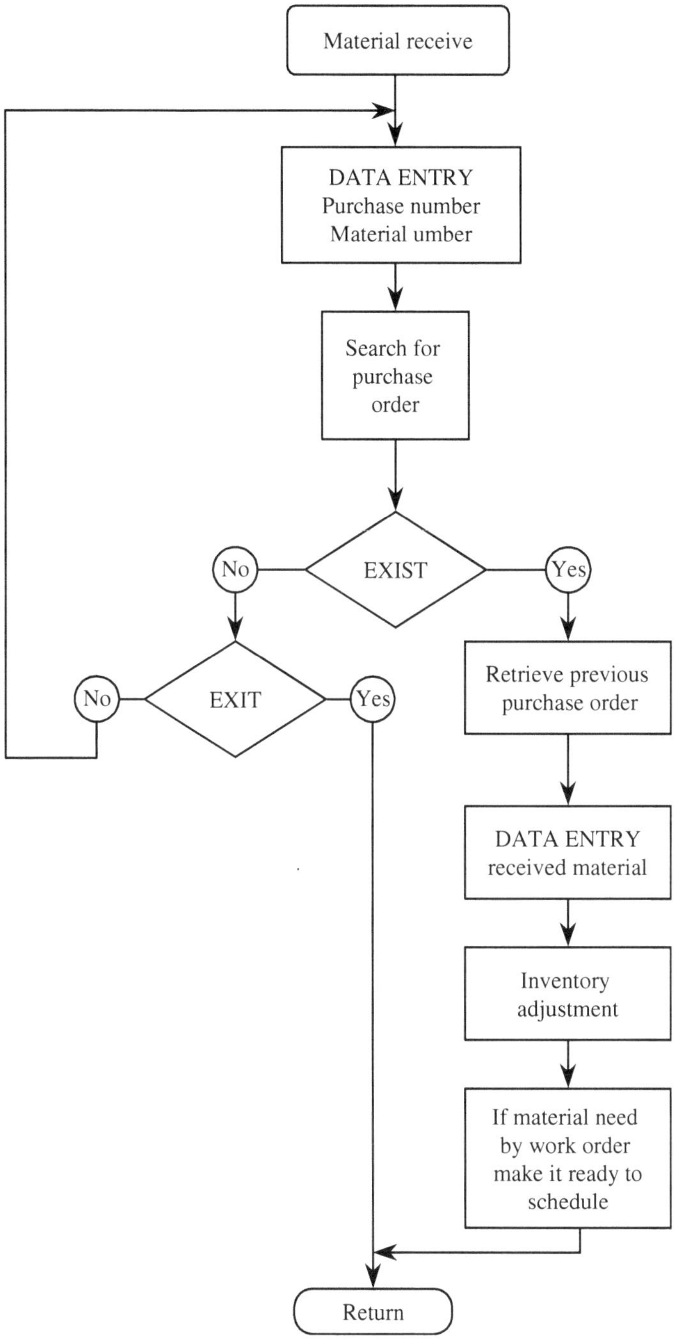

Fig. 10.6 Flowchart for material received

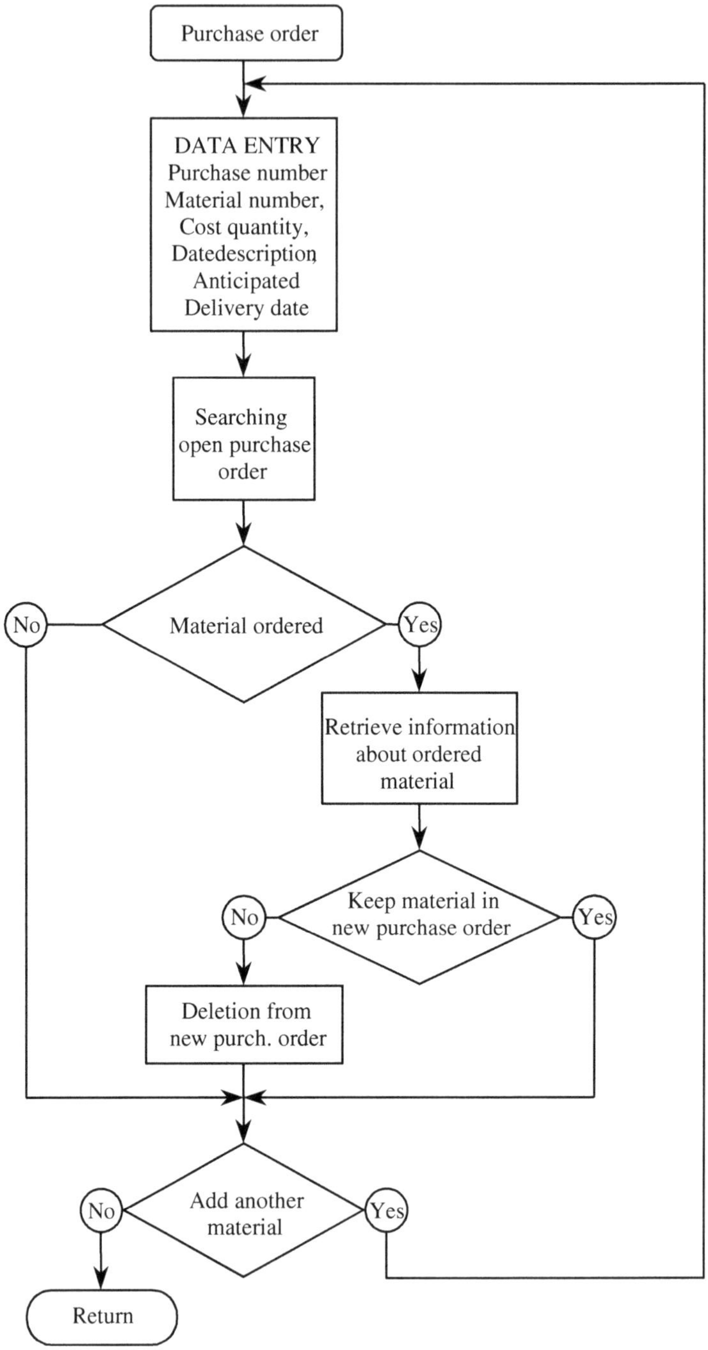

Fig. 10.7 Flowchart for purchase order

- Indicating material availability.
- Providing information on ordered material and their status.
- Deleting open purchase orders.
- Performing inventory adjustments.
- Initiating purchase orders.
- Searching for work orders waiting for material and change their status upon material arrival.

10.3.5 Performance Reporting Module

This module of CMMS interacts with all other modules to monitor the maintenance activities and provides various kinds of cost and performance reports. The module can be customized to generate all needed reports. However, the following reports are usually provided.

1. Cost reports. The system provides details about the maintenance costs. The costs can be prepared by equipment or by cost center where the equipment is located. The cost includes labor, spare parts, material, and facility costs.
2. Completed work orders. This is a summary of work orders completed in a specified time window.
3. Backlog report. This is a summary of work orders in the maintenance backlog. The work orders can be classified according to the cause of their backlog, which could be spare parts, crafts, or other technical reasons.
4. Work order status report. This includes the progress of all work orders.
5. Craft performance report. This report includes a summary of craft productivity.
6. Distribution of maintenance work by priority.
7. Estimated versus actual hours report.
8. Plant availability report. This report provides information about equipment availability.
9. Plant reliability report. This report provides information on major equipment reliability and mean time between failure and
10. Quality reports. This report provides information about repeat and substandard jobs and training records.

Other reports can be generated by using the system. These reports include work order delays, suspended jobs, and trends in utilization of crafts and others.

The output of the CMMS should be utilized for improvement purposes. The reports generated by the CMMS can be a useful input for the monthly maintenance report on any deficiency analysis.

10.3.6 Maintenance Reporting

The CMMS facilitates the preparation of reports to management. These reports aid management in taking actions to improve the status of the maintenance process.

The reports should be based on actual data and should be presented in summarized and graphical forms. Statistical indicators and trend analysis must be an integral part of such reports. The following general principal can be used for evaluating the quality of reporting:

1. The report should be concise and based on actual data. Standard cost figures and budgets should be used to form basic benchmarks.
2. Some periodic reports should be reported in a graphical form and must be compared with previous periods to show trends.
3. Important information such as equipment downtime, equipment losses, percentage of scheduled maintenance work, backlog reports, and maintenance quality reports should be available.
4. The CMMS should be able to provide most of these reports.
5. These should be prepared in close coordination with concerned departments, which should also provide interpretation of the reports' content.
6. Maintenance management should provide feedback on the reports, and action must be taken to correct any mishaps.

The maintenance departments should have a regular reporting system. Most maintenance departments have what is called the monthly maintenance report. The content of such a report should reflect the status of maintenance and typically include the following:

1. An abstract summarizing the main accomplishments and problems.
2. Summary of accomplished maintenance work. It should be a summary of work orders accomplished in terms of number and man-hours and be presented preferably in a tabular or graphical form and compared to the comparable figures from last three months to show trends.
3. Maintenance costs listed by cost centers or major equipment and compared with the last three months to identify trends.
4. Backlog reports in terms of man-hours presented in a graph or table and compared with the figures for last three months. Also, the backlog must be classified according to causes.
5. PM inspection report. This lists all the work orders initiated for preventive maintenance. It must include delayed PM work.
6. Downtime of major equipment.
7. Percentage of scheduled maintenance shifted to unscheduled status.
8. Open job report. The weekly status of all open work orders.
9. Efficiency reports comparing estimated hours versus actual.
10. Material and supply report. A brief description of material and spare part consumption and cost.
11. Monthly maintenance budget and variance.
12. The monthly maintenance report should be prepared prior to the end of the fifth day of the next month and submitted to the top executive responsible directly for maintenance, e.g., a director of maintenance.

13. The top executive, e.g., director of maintenance, should provide feedback on the report prior to the 20th day of the next month. It is suggested that the feedback should be in the form of a meeting with all maintenance managers to discuss the salient feature of the report and recommend possible improvement actions if necessary.

10.4 Evaluation of Computerized Maintenance Management Systems

The procedure in this section can be used for evaluating "off the shelf" computerized maintenance management systems available in the market or used as a guideline for developing an up-to-date CMMS. It has been reported in [2] that there are over 200 systems available in the market. The selection of a suitable one for an organization is not an easy task. The framework given in this section depends on the functional capability of the system. The basic functions of the system given in Fig. 10.6 will be used as a yardstick.

Let $i = 1, \ldots, n$ be the specific requirements or functions desired from a computerized maintenance management system for its intended utilization. These functions are work order management, preventive maintenance management, craftsmen management, inventory management, maintenance management reports, vendor evaluation, quality management, and general considerations [5, 6]. Let w_1, w_2, …, w_n be the respective weights or relative importance of these functions according to their intended use. For each of these n functions, let $j = 1, \ldots, m$ be the major characteristics which the system should offer for its effective use. The availability or effectiveness of these characteristics may vary from one system to another. These characteristics of a function are not equally important for every user. Let b_1, b_2, …, b_m be the relative weights of these m characteristics. In order to compare or select an appropriate CMMS from those available, assign a score out of ten to each of the characteristics based upon its availability through perfectness. Let S_{ij} be the score out of a possible ten score for characteristic i of function j.

Then, the total score for a function can be calculated as

$$A_i = W_i \sum_{j=1}^{m} b_j S_{ij}$$

The composite score for a CMMS is

$$\mathrm{CS} = \sum_{i=1}^{n} A_i = \sum_{i=1}^{n} \left[W_i \sum_{j=1}^{m} b_j S_{ij} \right]$$

The composite score (CS) can be calculated for all the available maintenance management systems, and the one with the highest CS is be preferred over the others.

The weights for each function W_i or the relative importance of each function can be determined by using the Analytic Hierarchy Process (AHP) method or some other statistically based method such as Delphi [3]. The AHP method has been explained and used for weight determination in Chap. 12. Table 10.2 provides suggested values for the weight of each function. Also, Tables 10.4, 10.5, 10.6,

Table 10.4 Preventive maintenance management (PM)

No.	Characteristics of the system	Grade
1.	The system will schedule PM by calendar date	
2.	Allows for more than one PM order per piece of equipment	
3.	Prints individual PM work orders	
4.	Provides a detailed description of the PM tasks to be performed	
5.	Prints a PM workload forecast for any given week or weeks	
6.	Allows for lead or lag time for scheduling the PM work order	
7.	Allows for a detailed listing of the PM tasks to be performed	
8.	Produces a report of overdue PM work orders	
9.	Projects the impact of the PM workload on the weekly schedule	
10.	Produces a report of the PM inspection results	
	Calculate total score $A_2 = W_2 \sum_{i=1}^{10} S_{2j}$	

Table 10.5 Inventory management

No.	Characteristics of the system	Grade
1.	Produces an inventory reorder report	
2.	Maintains unit price information for all spares	
3.	Identifies bin location of all spares	
4.	Produces a report for all work orders waiting for material	
5.	Attaches all material costs to the work order	
6.	Keeps a history record on all store items used	
7.	Keeps the economic order quantity for stock reorder	
8.	Keeps the max-min stock quantities on record	
9.	Produces a cost of inventory-on-hand report	
10.	Produces a complete store stock catalogue	
11.	Provides online parts inventory information	
12.	Allows for entering unused materials back into the stores inventory	
	Calculate total score $A_3 = W_3 \sum_{i=1}^{10} S_{3j}$	

Table 10.6 Quality control management

No.	Characteristics of the system	Grade
1.	Statistical techniques used are available	
2.	System keeps quality training records	
3.	System keeps the quality records	
4.	System keeps customer complaints record	
5.	Quality failure data available for analysis	
6.	Capable of tracing back data pertaining to quality	
	Calculate total score $A_4 = W_4 \sum_{i=1}^{6} S_{4j}$	

Table 10.7 Maintenance reports

No.	Characteristics of the systems	Grade
1.	Produces daily control reports	
2.	Produces equipment history report	
3.	Produces management reports on a monthly basis	
4.	Can produce management reports on demand	
5.	Produces reports tracking the system's backlog by craft	
6.	Produces failure analysis reports	
7.	Produces craft usage reports	
8.	Produces budget overrun reports	
9.	Reports on all incomplete work orders by priority	
10.	Capable of performing statistical and trend analysis	
	Calculate total score $A_5 = W_5 \sum_{i=1}^{10} S_{5j}$	

10.7, 10.8 and 10.9 provide the necessary function expected from a CMMS. Each table specifies the relevant characteristics of each function. The tables can be easily used to compute the composite score for any CMMS. These tables provide a framework and can be modified to suit different organization (Fig. 10.8).

10.5 Evolution of CMMs for Effective Maintenance

The elements of an effective CMMS are given in Sect. 10.2. Also, a framework for evaluating existing CMMS has been outlined in the previous section. The capabilities of CMMS system need to be enhanced and expanded to reflect recent developments in mathematical and statistical modeling. The following characteristics need to be added to the elements of CMMS.

Table 10.8 Vendor and service evaluation system

No.	Characteristics of the systems	Grade
1.	The vendor can supply installation support	
2.	The vendor has a documented installation program	
3.	The vendor will provide a list of installation references	
4.	The vendor will provide guidance during the data input	
5.	The vendor has a maintenance consultant staff to provide assistance in formatting data for entry into the system	
6.	The vendor provides documentation for installation, user manuals, and training manuals	
7.	The software can be self-installed	
8.	The vendor can provide training on-site or at their facilities	
9.	The vendor offers a planned enhancement and supports program for existing and future software	
	Calculate total score $A_2 = W_6 \sum_{i=1}^{10} S_{6j}$	

Table 10.9 General considerations

No.	Characteristics of the system	Grade
1.	The system is user-friendly	
2.	The system is menu-driven	
3.	The system is on-line and integrated	
4.	The system has an ongoing support program	
5.	The system keeps historical records until they are deleted from the system	
6.	The system has security password or code protection	
7.	The system runs on hardware already on-site	
8.	The system requires the purchase of special hardware	
	Calculate total score $A_7 = W_7 \sum_{i=1}^{10} S_{7j}$	

1. Effective control over maintenance scheduling by including heuristic scheduling rules, heuristic algorithms, and mathematical programming techniques.
2. Modeling characteristics, such as simulation of the maintenance systems.
3. Decision-making and forecasting capabilities.

The following modules of the CMMS need to be expanded in order to incorporate the above 3 characteristics.

1. Equipment management.
2. Work order control.
3. Performance reporting.

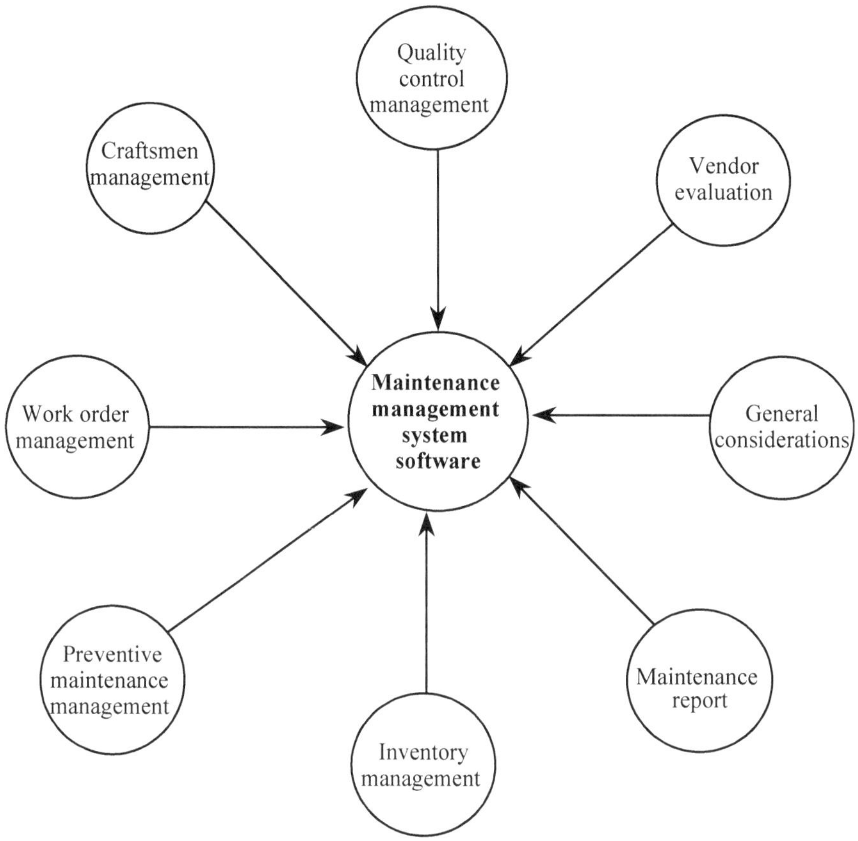

Fig. 10.8 Functions of computerized maintenance management systems

The equipment management module should include data on patterns of job arrivals, equipment history, reliability models, and predictive maintenance routines. Also, various statistical packages (as an example STATGRAPHIC [9]) must be an integral part of the module in order to develop needed statistical distribution for the stochastic programming and simulation models.

The work order control must be expanded to provide data needed for different scheduling model and stochastic simulation. The data include information about job standards and standard jobs. This process requires continuous monitoring due to the dynamic factors involved in the process. Scheduling model should be a part of the work order control subsystem. Therefore, the work order control module must include matrix and report generators in addition to optimization subroutines such Linear Interactive and Discrete Optimizer (LINDO), Generalized Interactive Nonlinear Optimizer (GINO) [4], and Optimization Software Library (OSL) [8].

Performance measures need to be expanded to include measures that relate the effect of maintenance policies on production and services. We propose that simulation models to be a part of the performance measures subsystem. This subsystem must include a special purpose simulation language such as Simulation Language for Alternative Modeling (SLAM), SIMSCRIPTS, or SIMON [5].

The statistical routines will provide forecasting capabilities and distribution function estimation which will be used in the stochastic model. The matrix generator will generate the optimization model and submit it to the optimizer which provides the optimal schedules. Then, the report generator will take the output of the optimizer and develop a schedule understandable to the foremen. The CMMS can be expanded to include decision-making capabilities. An example of this is the ability to stop the execution of low priority jobs upon the arrival of emergency jobs. Also, it should have the capabilities of revising job priorities depending on the past history.

If the suggested capabilities of the last paragraph are incorporated in the maintenance management information system, it becomes more like a decision support system (DSS) or an expert system (ES).

10.6 Summary

In this chapter, the basic elements of a CMMS have been discussed. Flowcharts are given presenting the components of each module in the system. The CMMS success is measured by its ability to support maintenance and provide timely information for effective decision making. The CMMS system must have most of the key indices for maintenance performance in order to quantify the status of maintenance.

The CMMS output will aid in developing the monthly maintenance report on which management should provide immediate feedback.

In addition, a systematic procedure is outlined for evaluating CMMS and future directions, and trends for CMMS development are described.

Exercises

1. What is a computerized maintenance management system (CMMS)?
2. List the possible modules in a computerized maintenance management system.
3. What are the functions of the work order control module?
4. What is the role of the equipment management module?
5. What is the information that can be obtained from the equipment management module.
6. How should the cost of maintenance be reported?
7. Select an organization and use Analytic Hierarchy Process (AHP) to determine the weight given to each function provided by a computerized maintenance management system.

8. Select an organization near you and evaluate their computerized maintenance management system.
9. Identify the training needs for an organization that would like to implement a CMMS similar to the one described in this chapter.

References

1. Bagadia k (1987) Microcomputer-aided maintenance management. In: Hartman E (ed) Industrial Engineering and Management Press, Institute of Industrial Engineering, Atlanta, Georgia, USA
2. Campbell JD (1995) Uptime: strategies for excellence in maintenance management. Productivity Press, Portland
3. Dalkay N, Helmers O (1963) An experimental application of the Delphi method to the use of the experts. Manage Sci 9(3):458–467
4. Duffuaa SO, Al-Sultan KS (1997) Mathematical programming approaches for the management of maintenance planning and scheduling. J Qual Maintenance Eng 3(3):163–176
5. Duffuaa SO, Andijani AA (1996) A stochastic simulation: an effective tool for maintenance, Working Paper, Department of Systems Engineering, King Fahd University of Petroleum and Minerals, Dhahran, Saudi Arabia
6. Niebel BW (1994) Engineering maintenance management, 2nd edn. CRC Press, Boca Raton (Marcel Dekker)
7. Raouf A, Ali Z, Duffuaa SO (1993) Evaluating a computerized maintenance management system. Int J Oper Prod Manage 13(3):38–48
8. Raouf A, Al-Jumah YA, Duffuaa SO (1990) A personal computer-based maintenance management expert system for hospitals. J Med Syst 14(3):119–128
9. STATGRAPHICS (1993) Corp., Statgraphics + Version 7
10. Wireman T (1986) Computerized maintenance management system. Industrial Press, New York

Chapter 11
Reliability-Centered Maintenance

11.1 Introduction

In the 1960s, the airline industry in the USA depended heavily on time-based preventive maintenance due to the belief that every component of a complex system has an age where a complete overhaul is needed to ensure safety and reliability. Due to that there was a hypothesis that the airline is performing excessive maintenance that is not economically viable and may be not needed. This hypothesis led to the launching of a major study to validate the failure characteristics of aircraft components. The study resulted in what became the Handbook for the Maintenance Evaluation and Program Development for the Boeing 747, more commonly known as MSG-1 (Maintenance Steering Group 1). MSG-1 was subsequently improved and became MSG-2 and was used for the certification of DC 10 and L 1011. In 1979, the Air Transport Association (ATA) reviewed MSG-2 to incorporate further developments in preventive maintenance; this resulted in MSG-3, the Airline/Manufacturers Maintenance Program Planning Document applied subsequently to Boeing 757.

Furthermore, United States Federal Aviation Administration (FAA) commissioned United Airlines to undertake a study of the effectiveness of time-based overhauls of complex components in the equipment systems of civilian jet aircraft. There was a belief that these time-based overhauls did little to reduce the frequency of failure and were uneconomical. This study was conducted at a time when wide-bodied aircraft were being designed, and the complexity of the equipment systems and their components was increased dramatically over that of prior designs. The key conclusion was that time-based overhauls of complex equipment did not significantly affect positively or negatively the frequency of failure. In some equipment, the frequency of failure was actually higher immediately following the overhaul. This study showed that the so-called bathtub *conditional probability of failure versus age* curve was only one of six major failure patterns. The most common failure pattern in complex equipment is one which shows a high "infant

S.O. Duffuaa and A. Raouf, *Planning and Control of Maintenance Systems*,
DOI 10.1007/978-3-319-19803-3_11

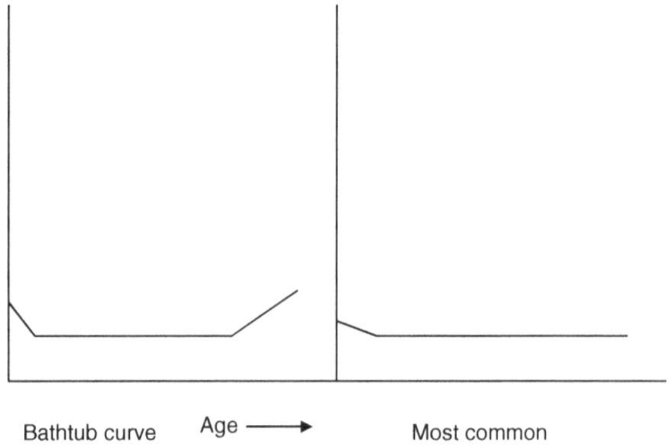

Fig. 11.1 Conditional probability of failure versus age

mortality," that is, the highest conditional probability of failure occurs in the first few periods of the equipment's age and then diminishes to a constant rate of failure, thereafter, as described in Fig. 11.1.

Scheduled, time-based overhauls "reset" the age back to zero, thereby increasing the probability of failure. For most of the life of complex equipment, failure is related to random events, such as shock loading, electrical surges, poor lubrication practices, and improper operation. These random events cause an accelerated deterioration of the equipment's performance, which often can be monitored using condition-based predictive maintenance techniques. This study leads to the birth of reliability-centered maintenance (RCM).

Reliability-centered maintenance is a logical methodology derived from this research in the aviation sector and uses the failure mode, effect, and criticality analysis (FMECA) tool. RCM is a process used to identify the most applicable and effective maintenance action(s) to ensure the highest practical standard of operating performance of a system or a component.

The purpose of this chapter is to present RCM as a viable approach for optimizing maintenance of systems by having an optimal mix of run to failure, time-based, condition-based, and design modification maintenance tasks. Section 11.2 outlines RCM principles and benefits followed the steps of RCM in Sect. 11.3. Section 11.4 outlines failure and its nature followed by RCM steps. Section 11.5 contains RCM implementation, and Sect. 11.6 provides a summary of the chapter.

11.2 RCM Goals, Principles, and Benefits

The success of RCM comes from the fact that it has clear goals, sound principles, and effective strategies. The main goal was to determine the optimal maintenance program using various maintenance strategies. RCM has principles that distinguish it from all other methodologies for developing maintenance programs. Some authors refer to these principles as RCM philosophy. In this section, RCM goals, principles, and benefits are presented.

11.2.1 RCM Goals and Objectives

RCM has the following goals:

- To determine the most cost-effective and applicable maintenance tasks to minimize the risk and impact of failure on systems/equipment function.
- Ensure high safety and reliability performance.
- Maintain system and equipment functionality in the most economical manner.

Specific RCM objectives as stated by Nowlan and Heap [6] are as follows:

- To ensure realization of the inherent safety and reliability levels of the equipment.
- To restore the equipment to these inherent levels when deterioration occurs.
- To obtain the information necessary for design improvement of those items where their inherent reliability proves to be inadequate.
- To accomplish these goals at a minimum total cost, including maintenance costs, support costs, and economic consequences of operational failures.

The above goals and specific objectives are clear derive for effective maintenance programs that usually result from the application of RCM methodology.

11.2.2 RCM Principles

RCM has the following key principles that distinguish it from other methodologies for maintenance:

- Preservation system of equipment function: The focus here is to keep the system performing its function not to keep it operating as though it is new. This tells us that as far as the system performing its function, there is no need for excessive maintenance which may cause failure in some cases. This principle has led to a reduction in time-based preventive maintenance in the airline industry that reduced cost and improved reliability of systems. Redundancy of function through multiple equipment improves functional reliability, but increases life cycle cost in terms of procurement and operating costs.
- Focus on systems: RCM focuses on systems than component, since functions are usually driven by systems.
- RCM is reliability-centered: It treats failure statistics as it relates to age. It seeks to know the conditional probability of failure at specific ages (the probability that failure will occur in each given operating age bracket).
- Safety and economics are the key criteria: Safety must be ensured first at any cost; followed by costs that result from the impact on production and operation.

- Design limitations exist: RCM objective is to maintain the inherent reliability of the equipment design, recognizing that changes in inherent reliability arises from design rather than maintenance. Maintenance can, at best, only achieve and maintain the level of reliability for a system.
- Feedback is necessary for improvement: RCM recognizes that maintenance feedback can improve on the original design. In addition, RCM recognizes that a difference often exists between the perceived design life and the intrinsic or actual design life.
- Failure is any unsatisfactory condition: failure may be either a loss of function (operation ceases) or a loss of acceptable quality (operation continues).
- Maintenance tasks should be derived based on logic: RCM uses a logic tree to develop and screen maintenance tasks.

11.2.3 RCM Benefits

The primary driving force behind the invention of RCM is the need to develop a maintenance strategy that can adequately address the systems availability and safety without creating a totally impractical cost requirement. RCM benefits include the following [5, 7]:

- Increase in Plant availability.
- Survey conducted by Electric Power Research Institute provided more evidence that RCM has impact on cost.
- Plant trip reduction.
- Documented basis for preventive maintenance (PM).
- Efficient PM planning.
- Decrease in corrective maintenance.
- More accurate spare parts identification.

Smith [8] provided the following distribution in the type of maintenance strategies to demonstrate the impact of RCM brought in power plants maintenance (Table 11.1).

Table 11.1 Impact of RCM on maintenance task distribution

Type of maintenance tasks	Years: task distribution in %		
	1964	1969	1987
Time directed	58	31	9
Condition directed	40	37	40
Run to failure	2	32	51

11.3 RCM Methodology

RCM has a systematic methodology that consists of seven steps. The seven steps are as follows [3, 6, 8, 9]:

1. System selection and information collection;
2. System boundary definition;
3. System description and functional block diagram;
4. Functions and functional failure;
5. Failure mode and effective analysis (FMEA);
6. Logic decision tree analysis (LTA); and
7. Task selection.

In the next subsections, each step is explained and described.

11.3.1 System Selection and Information Collection

RCM is best presented and implemented at a system level due to the fact that functions are best captured at the system level. The component level lacks defining significance of functions and functional failure, while plant-level analysis makes the whole analysis intractable. The important question faced at this stage is which system should be selected? The following are criteria that guide the selection:

1. Systems with a high number of corrective maintenance tasks during recent years;
2. Systems with a high number of preventive maintenance tasks and or costs during recent years;
3. A combination of scheme 1 and 2;
4. System with a high cost of maintenance;
5. Systems contributing significantly toward plant outages/shutdowns (full or partial) during recent years;
6. Systems with high concern relating to safety; and
7. Systems with high concern relating to environment.

Past experience has shown that all of these criteria except scheme 6 and 7 yield more or less the same results. An indicator of a good selection is that systems chosen for RCM program results in a significant improvement over the current situation.

The next task, after selecting a system, is collecting information related to the selected system. A good practice is to start collecting key information and document right at the onset of the process. The following are documents that may be required in a typical RCM study:

- P&ID (piping and instrumentation) diagram;
- Systems schematic and/or block diagram;

- Functional block diagram;
- Equipment design specification and operations manuals (a source of finding design specifications and operating condition details);
- Equipment history file (failure and maintenance history in specific);
- Other identified sources of information, unique to the plant or organizational structure. Example includes industry data for similar systems; and
- Current maintenance program used for the system. This information is generally not recommended to collect before step 7, in order to avoid biases that may affect the RCM process.

11.3.2 System Boundary Definition

System boundary definition is needed for the following reasons:

- It provides an exact knowledge of what is included and not included in a system in order to make sure that any key system function or equipment is not neglected (or not overlapped from another system). This is especially important if two adjacent systems are selected.
- Boundary definition also includes system interfaces (both IN and OUT interfaces) and interactions that establish inputs and outputs of a system. An accurate definition of IN and OUT interfaces is a precondition to fulfill step 3 and 4 below.

There are no clear rules to specify a system boundaries; however, as a general guideline, a system has one or two main functions with a few supporting functions that would make up a logical grouping of equipment. However the boundary is identified, there must be clear documentation as part of a successful process.

11.3.3 System Description and Functional Block Diagram

This step is important and will set the stage for a successful RCM process. The step has the following five elements:

- System description;
- Functional block diagram;
- In/out interfaces;
- System work breakdown structure; and
- Equipment history.

This step generally involves a form that documents baseline characterization of a system which is eventually expected to be used in stipulating PM tasks.

The five elements established during this step are as follows:

(i) *System description*:

Key elements of a system description include: Functional description/key parameters, redundancy features, protection features, key instrumentation features, and safety features. Figure 11.2 provides a typical form for system description.

There are tangible benefits of a good system description because it will ensure a comprehensive understanding of the system and help in identifying critical parameters that contribute to degradation and loss of function. It also identifies key design and operational parameters that directly affect the performance of the system function.

(ii) *Functional block diagram(FBD)*

The FBD is a top-level representation of the major functions the system performs. A functional block diagram describes a function between inputs and outputs and should describe the system in one diagram. The diagram consists of functions only, labeled functional subsystems, and in/out interfaces shown with arrows. To represent this functional behavior, FBD or functional block diagram (Fig. 11.3) is used to gather information prior to drawing the diagram.

RCM: System Analysis (System Description)		
Date:	Plant:	Location:
System Name: System ID: System Location:		RCM Analyst(s): 1. 2.
Functional Description		
Key Parameters		
Key equipment		
Redundancy Features		
Safety Features		

Fig. 11.2 Typical RCM system analysis form

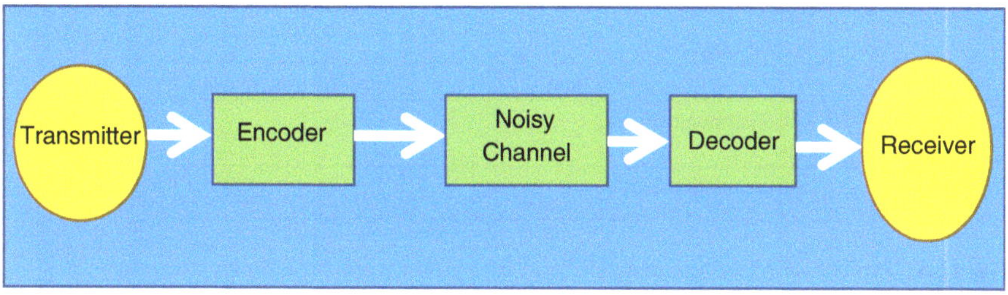

Fig. 11.3 Functional block diagram

Figure 11.3 shows an example of a functional block diagram for a communication system. The system consists of a transmitter, an encoder, a hoist channel, a decoder, and a receiver.

(iii) *In/out interfaces*

Systems have IN and OUT interfaces. IN interfaces exist within a system while OUT interfaces exist at the boundaries of the system, making themselves the principle objects to preserve system functions. A point to note is that the IN interfaces might be the OUT interfaces in some other systems. If an interface is within a system boundary connecting to system environment, it is called an internal OUT interface. Figure 11.4 shows a form used to document interfaces.

(iv) *Systems work breakdown structure*

Systems work breakdown structure or SWBS is a term used to identify a list of equipment/components for each of the functions shown in functional block diagram. This list is defined at the component level of assembly that resides with the system boundary. Identification of all components within a system is essential as otherwise it will eliminate these unlisted components out of the PM considerations. A typical SEBS form is shown in Fig. 11.5.

RCM: System Analysis (Interface Definition)		
Date:	Plant:	Location:
System Name: System ID: System Location:		RCM Analyst(s): 1. 2.
IN interfaces		
OUT interfaces		
Internal OUT interfaces		

Fig. 11.4 Typical RCM system analysis form for interface definition

RCM: System Analysis (System Work Breakdown Structure)		
Date:	Plant:	Location:
System Name: System ID: System Location:		RCM Analyst(s): 1. 2.
	Item	Number of item used
Non-instrumentation List		
Instrumentation List		

Fig. 11.5 Typical RCM system analysis form for system work breakdown structure

(v) *Equipment history*

Equipment history is presented in Chap. 6. It is the record of the equipment failure, failure mode, cause of failure, and maintenance action taken whether preventive or corrective in addition to other information. Any decent maintenance system must have equipment history files for its equipment. If it is not available, then it can be obtained from work orders used for corrective and preventive maintenance. This data can be recorded in a form as shown in Fig. 11.6.

11.3.4 System Functions and Functional Failure

The fourth step identifies and lists all system functions. As a guide for identifying functions, every out interface should be captured into a function statement and any internal out interfaces between functional subsystems can be a source for a function. An important point to note is that these statements are for defining system functions and not the equipment. With the definition of system functions comes the functional failures. In RCM, the focus is on functions and functional failures. The functional failures are more than just a single statement of loss of function. The loss conditions may be two or more (e.g., complete paralysis of the plant or major or minor deprivation of functionality). This distinction is important and will lead to the proper ranking of functions and functional failures. Figure 11.7 provides a form to document functions and functional failures. The following are the examples for the correct and wrong statement of functions:

- Provide 1500 psi safety relief valves (wrong statement because the statement is about equipment);
- Provide for pressure relief above 1500 psi (correct; the focus is on function);
- Provide a 1500 gallon per minute (gpm) centrifugal pump on the discharge side of header 26 (wrong); and
- Maintain a flow of 1500 gpm at the outlet of header 2

Fig. 11.6 Typical RCM system analysis form for equipment history

RCM: System Analysis (Equipment History)			
Date:	Plant:		Location:
System Name: System ID: System Location:			RCM Analyst(s): 1. 2.
Component	Date	Failure Mode	Failure Cause

RCM: System Analysis (Function and Functional Failures)	
Step 4: Function and functional failures	
Plant:	Plant ID:
System name:	System ID:
Analyst:	

Function no. functional	Functional failure no.	Functional or failure description

Fig. 11.7 Function and functional failure form

11.3.5 Failure Modes and Effects Analysis (FMEA)

Failure modes and effects analysis (FMEA) is a basic tool used in reliability engineering to assess the impact of failures. It is a systematic failure analysis technique that is used to identify the failure modes, their causes, and consequently their fallouts on the system function. FMEA analysis rates each potential failure mode and effect based on the following three factors:

- *Severity*—the consequence of the failure when it happens;
- *occurrence*—the probability or frequency of the failure occurring; and
- *detection*—the probability of the failure being detected before the impact of the effect is realized.

Then these three factors are combined in one number called the risk priority number (RPN) to reflect the priority of the failure modes identified. The risk priority number (RPN) is simply calculated by multiplying the severity rating, the occurrence probability rating, and the detection probability rating.

$$\begin{aligned}\text{RPN} &= (\text{severity rating}) * (\text{occurrence probability rating}) \\ &\quad * (\text{Detection Probability rating})\end{aligned}$$

FMEA process is usually documented using a matrix similar to the one shown in Fig. 11.8.

11.3.6 Logic or Decision Tree Analysis (LTA)

The purpose of the LTA is to prioritize the resources to be committed to each failure mode. The prioritization is based on the impact of the failure mode. RCM processes a simple and intuitive structure for this purpose. The structure utilizes two criteria, i.e., safety and cost, that arise from plant full outage. The LTA has three questions that enable a user, with minimal efforts, to place each failure mode into one of the six categories. Each question is answered as yes or no only. Each category (also known as a bin) forms natural segregation of items of respective importance. The LTA scheme is shown below in Fig. 11.9.

RCM: Systems Analysis (FEMA)						
Step 5: Failure mode and effect analysis					Ref no.::	Date:
Functional failure (FF) no.:			FF name:			
Plant:			Plant ID:			
System name:			System ID:			
Team:			Head:			
Component	Failure mode	Failure cause	Failure Effect			LTA
			Local	System	Plant	

Fig. 11.8 Failure mode and effect analysis (FMEA)

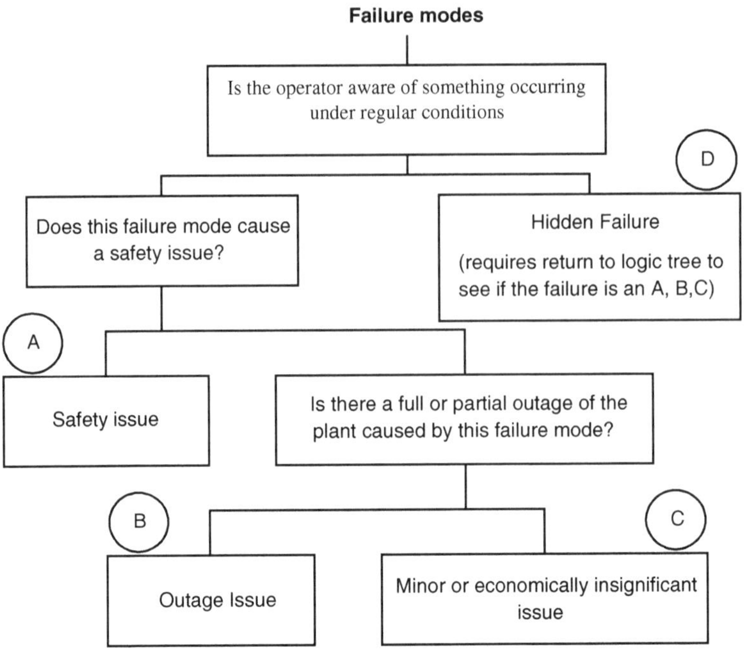

Fig. 11.9 Logic tree analysis

The six classification categories for the failures are A, B, C, D/A, D/B, or D/C. For the priority scheme, A and B have higher priority over C when it comes to allocation of scarce resources and A is given higher priority than B. In summary, the priority for PM task goes in the following order:

1. A or D/A;
2. B or D/B; and
3. C or D/C.

11.3.7 Task Selection

In this step, RCM methodology allocates PM tasks and resources. This is the stage where the maximum benefit from RCM may be obtained. The task selection process requires that each selected task must be applicable and effective. Here, "applicable" means that the task should be able to prevent failures, detect failures, or unearth hidden failures, while "effective" is related to the cost effectiveness of the alternative PM strategies. If no PM task is selected through the LTA, the only option is to run equipment to failure. This activity requires contribution from the maintenance personal as their experience is invaluable in the correct selection of the PM task. After selecting the tasks, the set of all run-to-failure (rtf) tasks are subjected to a final sanity check. The purpose of the check is to review critically all component failures that are treated as run-to-failure cases to see if this task is appropriate. If an rtf task fails any of the following tests or creates a conflict, the PM or the current task is kept. The following are the checks:

- Marginal effectiveness: It is not clear that the rtf costs are significantly less than the current PM costs.
- High-cost failure: While there is no loss of critical function, the failure mode is likely to cause extensive damage to the component that should be avoided.
- Secondary damage: Similar to the second item, except that there is a high probability extensive damage in neighboring components.
- OEM conflict: The original manufacturer recommends a PM task that is not supported by RCM. It is very sensitive if warranty conditions are involved.
- Internal conflict: Maintenance or operation feels strongly about the PM task that is not supported by RCM.
- Regulatory conflict: Regulatory body established the PM, such as the Environmental Protection Agency (EPA).
- Insurance conflict: similar to the above two.

11.4 RCM Standards

Two RCM methodologies are recognized. The classical methodology presented in Sect. 11.3 above is based on the work of Nowlan and Heap [8] or one of the commercial aviation standards such as MSG-3. The classical methodology requires the commitment of a core group of individuals and a facilitator for a few hours per week over a period of a month or two.

There is a perception that the classical methodology requires too much resources and that led to the development of what is known as light or rapid RCM methodologies. These methods typically analyze the existing maintenance tasks with the objective of adjusting intervals and eliminating unnecessary maintenance, where real RCM begins with the analysis of equipment functions, functional failures, and failure modes. The rapid methodologies may sometimes be quicker to implement than classical RCM. However, there is a real danger sometimes of failing to identify serious risks to production, safety, or environment.

The potential for confusion between “classical” and “light” methods was a key driver for the development of the US Society of Automotive Engineers (SAE) standard JA1011. This short document defines the requirements that any process must satisfy in order to be considered as RCM. A second standard, SAE JA1012, is longer and elaborates on the first document, including examples of possible RCM decision diagrams. There are other standards that define what constitutes a true RCM. These include as MSG-3, AP 100-C22, defense-related standards such as MIL HDBK 2173 or the UK DEF-STAN 00-45. It is advisable not to hang up on which standard to use and focus on the one that serves the organization's need. The SAE standards are based on the following seven questions that align very well with the seven steps presented in Sect. 11.3:

1. What is the item supposed to do and its associated performance standards?
2. In what ways can it fail to provide the required functions?
3. What are the events that cause each failure?
4. What happens when each failure occurs?
5. In what way does each failure matter?
6. What systematic task can be performed proactively to prevent, or to diminish to a satisfactory degree, the consequences of the failure?
7. What must be done if a suitable preventive task cannot be found?

11.5 RCM Implementation

RCM implementation can be viewed as a process with four stages [1, 2, 4, 7]. Each stage consists of a number of tasks that must be executed in order to ensure successful implementation. The four stages are as follows:

- Stage 1: Planning and organizing for RCM.
- Stage 2: Analysis and design.
- Stage 3: Scheduling and execution.
- Stage 4: Assessment and improvement.

The following subsections present the process for RCM implementation.

11.5.1 Planning and Organizing for RCM

This stage is important for the success of RCM implementation. Top maintenance management must seek organization commitment for the RCM project and ensure the needed resources are provided. In some situations, it may be better to conduct a pilot RCM project. The following must be addressed carefully:

1. Organization: Formation of the RCM team and selection of a facilitator. The facilitator must be knowledge about the RCM process. The team must include experienced people in the areas that will be impacted by the application of the RCM. The team must use a clear system for reporting progress and challenges.
2. Training: A training program on RCM should be conducted at different levels. Management should be provided an awareness program about RCM and its benefit. A well-structured training program on RCM methodology and implementation should be provided to the team.
3. Avail resources: Estimate what type of resources is needed and ensure their availability.

4. Manage expectations: Establish a baseline for the current performance and the expected benefits from RCM.
5. Schedule: Prepare a schedule for RCM project on a Gantt chart with the necessary resources. A schedule will facilitate follow-up and monitoring.
6. Change management program: Develop a change management program to mitigate resistance to the RCM project and ensure buying. The program may have an awareness program, training, and reorganization.

11.5.2 Analysis and Design

This stage deals with the development of the optimized maintenance program using RCM methodology. It includes the selection of the methodology and its application. The team is expected to follow the RCM steps. For each maintenance task design its maintenance procedure. The procedure should specify task requirements in terms of manpower, spare parts, standard time, and schedule. The outcomes of this stage are maintenance tasks for each functional failure.

11.5.3 Execution Stage

In this stage, the team integrates RCM tasks in the maintenance schedule. The impact of the new tasks is estimated in terms of benefits and cost. The change management program is reviewed and enhanced if needed.

11.5.4 Assessment and Feedback Stage

At this stage, the RCM project is implemented either in full or as a pilot case. The impact of the RCM project must be measured and assessed using relevant realistic performance measures. The measures may include availability, quality rate, reliability, and cost. Comparisons with prior RCM performance and current target will reveal RCM impact and aide in identifying needed improvement.

11.6 Summary

Reliability-centered maintenance is the process of determining the most effective maintenance program. RCM seeks to optimize the preventive maintenance program. RCM started in the airline industry and has been implemented in defense, nuclear, power, and petrochemical industries. RCM reduces unnecessary maintenance, improves reliability, and reduces costs. This chapter introduced RCM and its methodology. The methodology is a systematic approach that consists of seven steps. Guidelines for RCM implementation are also provided.

Exercises

1. Define reliability-centered maintenance (RCM).
2. List the goals and the objectives of RCM.
3. What are the main principles of RCM?
4. State the steps of RCM.
5. Make a mapping between the RCM seven steps and RCM questions to demonstrate their alignment.
6. How are the system functions determined?
7. What is a functional failure?
8. What are the criteria for system selection?
9. What are the main criteria for task selection?
10. What are the main roles and the outcomes of the logic tree analysis?
11. What is the rationale for the sanity check?
12. Describe the checks conducted in the sanity check.
13. Collect all the different standards for RCM and compare and contrast them.
14. Select a system in an organization in your area and apply RCM.
15. Explain how to establish a baseline for the current performance of the existing maintenance program.

References

1. Anderson RT, Neri L (1990) Reliability centered maintenance: management and engineering methods. Elsevier Applied Sciences, London
2. August J (1999) Applied reliability-centered maintenance. Pen Well, Oklahoma
3. http://www.numeratis.com/rcm/talktomanufacturers
4. Moubray J (1997) Reliability centered maintenance. Butterworths-Heineman, Oxford
5. NASA Reliability Centered Maintenance Guide for Facilities and Collateral Equipment, 3–24 February 2000
6. Nowlan FS, Heap HF (1978) Reliability centered maintenance, United Airlines and Dolby Press, sponsored and published by the Office of Assistant Secretary of Defense
7. Siddiqui AW, Ben Daya M (2009) Reliability centered maintenance. In: Ben-Day M, Duffuaa S, Raouf A, Knezevic J, Ait-Kadi D (eds) Handbook of maintenance management and engineering, edited by. Springer, London
8. Smith AM, Hinchcliffe GR RCM Gateway to world class maintenance. Elsevier Butterworth-Heinemann, Burlington
9. The Collinson Corporation Introduction to reliability-centered. reprinted courtesy of: P.O. Box 495578 Garland, Texas 75049–5578

Chapter 12
Total Productive Maintenance

12.1 Introduction

Nakajima, who is considered by many in the literature as the father of total productive maintenance (TPM), defines it [5] as "*productive maintenance carried out by all employees through small group activities.*" He also adds "TPM is equipment maintenance performed on a company wide basis." The authors define TPM as a management approach to maintenance that imports total quality management (TQM) philosophy and techniques to maintenance. TPM focuses on involving all employees in the organization in equipment improvement. This approach has its origins at Nippondenso, a subsidiary of the Toyota Motor Company in the 1960s and evolved in the Japanese manufacturing sector beginning with the application of American- and European-style preventive maintenance, and progressing to the application of TQM and just-in-time manufacturing concepts to the equipment maintenance arena. In 1971, the first prize is awarded by Japan Institute of Plant Engineers (JIPE) for successful implementation of TPM and still going on.

TPM started in Japan and has spread to the Far East, Europe, South America, and the USA. It became a recognized methodology for improving equipment reliability and plant productivity. TPM eliminates losses resulting from unplanned downtime, reduced speed, and quality. TPM brought to industry overall equipment effectiveness (OEE) as a measure that combines losses from unplanned downtime, reduced speed, and quality. The purpose of this chapter is to present the concepts and implementation of TPM. Section 12.2 outlines the goals and the key elements of TPM followed by autonomous maintenance in Sect. 12.3. Section 12.4 presents equipment management, and Sect. 12.5 describes TPM implementation. Section 12.6 provides a brief summary of this chapter.

S.O. Duffuaa and A. Raouf, *Planning and Control of Maintenance Systems*,
DOI 10.1007/978-3-319-19803-3_12

12.2 TPM Goals and Key Elements

The JIPE in its definition of TPM in 1971 stated that TPM seeks the following five key goals:

- maximize OEE, which includes availability, process efficiency, and product quality;
- take a systematic approach to reliability, maintainability, and life cycle costs (LCC);
- involve operations, materials management, maintenance, engineering, and administration in equipment management;
- involve all levels of management and workers; and
- improve equipment performance through small group activities and team performance.

The key elements of TPM include the following:

- **Autonomous maintenance**: Equipment operators are the focal point of TPM activities. Although most operators understand what their equipment does, few understand the underlying mechanisms of how it does it. The term "autonomous maintenance" is used to describe the activities of the operators, which relate to equipment maintenance, and to the independent study nature of the other equipment improvement activities. Operators would perform cleaning, inspection, lubrication, adjustments, and minor component change outs and other light maintenance tasks requiring some training and instruction, but not comprehensive craftsman skills. The operator gradually learns how to diagnose equipment problems before they become serious.
- **Equipment Management**: In TPM, whenever equipment performs at a level less than is required, the performance loss is recorded and monitored. These losses can be grouped into six categories: breakdowns, setup and adjustments, idling and minor stoppages, reduced speed, defects, and yield losses. Breakdowns and setups cause downtime and impact availability, reduced speed impacts the cycle time and defects, and yield losses impact quality. OEE is the key TPM performance measure and is the product of availability, cycle time, and quality rate. The operator and maintainer are trained to identify problems related to OEE and perform root cause analyses in teams to investigate the losses.
- **Systematic Planning and Continuous Improvement**: Within the maintenance department, the TPM methodology encourages the development of systematic planning and control of preventive and corrective maintenance, and fully supports the autonomous activities performed by the operator. In plants where the basic operating and maintaining environment has been improved to the point of diminishing returns, active maintenance prevention activities are undertaken, as described earlier in the sections on designing for maintainability. Throughout, there should be a strong emphasis on improving operator and maintainer skills. Spending on training is customarily on the order of 5–8 % of the labor budget.

12.3 Autonomous Maintenance

The benefits of involving the operators in the success of TPM cannot be overemphasized. A pragmatic way of achieving this is by using a systematic, data-based approach to skill transfer. Skill transfer is the process of moving tasks requiring lower skills from the exclusive domain of one work group to a shared task zone. Under this policy, an operator who has been properly trained and certified can perform a mechanic's task and vice versa. This partnership between operations and maintenance integrates maintenance and operation/manufacturing and has many benefits that include the following:

- Operators and mechanics become multi-skilled, which leads to job enrichment and improved flexibility of workers.
- The involvement of operators in routine maintenance builds a sense of responsibility, pride, and ownership.
- Delay times are reduced and productivity is increased.
- Teamwork between operations and maintenance is promoted.

12.4 Equipment Management

Equipment is the focus of TPM. Well-maintained equipment reduces losses resulting from unavailability, low speed, and quality defects. This effort for effective equipment management starts by identifying the major losses of equipment. As noted above, the following six losses limit equipment effectiveness:

1. Equipment failure (breakdown),
2. Setup and adjustment downtime,
3. Idling and minor stoppages,
4. Reduced speed,
5. Process defects, and
6. Reduced yield.

The ultimate goal of TPM with respect to equipment management is to increase its effectiveness to its highest potential and to maintain it at that level. This can be achieved by understanding the above losses and devising means of eliminating them.

Equipment failures
Breakdowns account for a large percentage of total losses. Every attempt should be made to avoid them. In order to maximize equipment effectiveness, breakdowns must be reduced to zero by changing the attitude that breakdowns are inevitable.

Setup and adjustment downtime
When production of a given product type ends and the equipment is adjusted to get ready to produce another product type, there are losses due to set up downtime and defective products. These losses can be reduced by reducing setup time. Many companies are working to achieve single-minute setups.

Idling and minor stoppages
Production may be interrupted because of a malfunction or a production machine being idle between products. The sources of these losses must be identified and eliminated. The elimination of minor stoppages is an essential precondition for automated production.

Reduced speed
These losses correspond to the difference between equipment design speed and actual operating speed. Reduced speed may be due to mechanical problems and defective quality or may be imposed by the operator in fear of abusing the equipment. In other cases, the optimal speed may not even be known. Productivity improvement results from increasing the speed if the reasons for operating at a reduced speed are identified and eliminated.

Process defects
These are losses in quality caused by the equipment. The conditions causing the defect must be identified and eliminated.

Reduced yield
These are the start-up losses occurring during the early stages of production from the beginning to its stabilization.

Improving equipment effectiveness requires that these losses be measured. The above-mentioned six losses affect equipment availability, efficiency, and the quality of the product as follows:

1. Equipment availability is affected by setup and adjustments, and equipment failures.
2. Equipment efficiency is affected by idling, minor stoppage, and reduced speed.
3. Reduced yield and process defects affect product quality.

The overall equipment effectiveness (OEE) is given as follows:

$$\text{OEE} = \text{Availability} \times \text{Performance efficiency} \times \text{Quality rate}$$

where

$$\text{Availability} = \frac{\text{Loading time} - \text{Downtime}}{\text{Loading time}}$$

loading time is defined as the available time minus planned downtime. Planned downtime refers to the downtime officially scheduled in the production plan, such as scheduled maintenance and management activities.

$$\text{Performance efficiency} = \frac{\text{Theoretical cycle time} \times \text{Amount processed}}{\text{Operating time}}$$

$$\text{Quality rate} = \frac{\text{Amount processed} - \text{Defective amount}}{\text{Amount processed}}$$

Overall equipment efficiency can be reduced to the following:

$$\text{OEE} = \frac{\text{Theoretical cycle time} \times \text{Amount processed} \times \text{Quality rate}}{\text{Loading time}}$$

Theoretical cycle time and loading time are constant per day. Therefore, OEE is directly related to the number of items of good quality. OEE can be improved by enhancing the availability, performance efficiency and, most importantly, the quality rate. It can be seen that availability can be improved by reducing downtime and that performance efficiency can be improved by reducing the cycle time.

In order to improve the quality rate, the state at which the equipment produces a high-quality product must be identified. Then, a maintenance policy should be established to keep the equipment in this state. Therefore, the goal of maintenance should be to keep the equipment in the state where zero defects are produced.

12.5 TPM Implementation

The practice indicates that TPM is usually implemented over a horizon of 3 years. Nakajima suggests four stages for implementing TPM. The stages are preparation, preliminary implementation, TPM implementation, and stabilization. Each stage consists of a number of steps. The total number of steps needed to implement TPM is 12. An excellent treatment of TPM implementation is provided by Nakajima [5].

12.5.1 Preparation Stage

This stage has the following five major steps:

- **Announce top management's decision to introduce TPM**: In this step, top management informs employees of the decision to implement TPM and shows enthusiasm and strong commitment.

- **Launch an educational campaign to introduce TPM**: At this step, promotion and training for TPM implementation must start immediately after the announcement. The education and training is designed to eliminate resistance and raise moral. The education should be tailored to the role that will be played by the group in TPM implementation. Retreats, seminars, and presentations are suitable for senior and middle management. Equipment management instruction is suitable for operators and maintenance staff.
- **Create organizations to promote TPM**: The organization structure is usually based on an organizational matrix consisting of horizontal groups and project teams at each level of the vertical management or organization. JIPE recommends a network of overlapping small groups organized at every level from top management to the work floor. Each group leader participates as a member in a small group at the next level. The group leader serves as a link between levels, facilitating vertical and horizontal communication.
- **Establish basic TPM policies**: TPM leadership establishes goals and basic polices for medium and long-range planning. For example, a basic policy could be "To reduce losses by eliminating breakdown, defects, and accidents while improving employee morale." The company should prepare a manual to be the charter for TPM implementation.
- **Formulate a master plan for TPM development**: The charter mentioned in the previous step must contain a master plan for TPM development. TPM development should be centered on the following basic five items:

  - Improving equipment effectiveness by eliminating the six big losses,
  - Establishing an autonomous (operator) maintenance program,
  - Quality assurance,
  - Establishing a schedule for planned maintenance, and
  - Education and training to enhance skills.

12.5.2 Preliminary Implementation

- **Hold TPM kickoff**: This is the first step in TPM implementation and is the start of the activities against the six big equipment losses. During the preparation stage, management and professional staff play the major role. At this step, everyone must participate and work to eliminate the big losses. In the meeting, management reports on accomplishment in the preparation stage including TPM promotion structure, basic TPM goals and policies and master plan for TPM development. Management affirms management commitment.

12.5.3 TPM Implementation

This stage consists of the following five steps:

- **Improve effectiveness of each piece of equipment**: This is the first step of the five steps of TPM development. Project teams consisting of engineers, maintenance staff, line supervisors, and small group members are formed to make improvements through the elimination of big losses. Some project teams may have doubts about the viability of TPM. To reduce or eliminate the doubts, TPM effectiveness can be demonstrated by focusing on equipment with high losses as a pilot. Root cause analysis and PM analysis developed by JIPE can help in making improvement. In PM analysis, the P stands for "problem" or "phenomenon" or "physical," while the M stands for "mechanism," "man," "machine," or "material."
- **Establish an autonomous maintenance program for operators**: This is the eighth step and the second of the five TPM development activities and a unique feature of TPM. Having autonomous maintenance requires cultural transformation where the concept of division of labor has to be overcome and everyone from top to bottom in the organization believes that it is feasible for operators to perform autonomous maintenance and be responsible for their equipment. Operators must be trained to carry out autonomous maintenance such as inspections, lubrication, cleaning, and other simple preventive maintenance tasks. Nakajima in his classical book on TPM [5] provides a detailed description for developing and perfecting autonomous maintenance.
- **Develop a scheduled maintenance program for the maintenance department**: This is the ninth step and is one of the activities for developing TPM. This step starts before completing autonomous maintenance. Existing planned and scheduled maintenance need to be evaluated and improved as part of TPM implementation. In order to optimize the preventive maintenance program, reliability-centered maintenance is an option to use according to Ben Daya [2].
- **Conduct training to improve operation and maintenance skills**: This is the tenth step in TPM implementation and the fourth step in TPM development activities. Training is critical for TPM implementation, and it is an investment in people to enable them to manage their equipment. Training includes maintenance techniques, basic equipment operation, testing, trouble shooting, and planning and scheduling.
- **Develop an effective management program**: This is the eleventh step and the last step in TPM development activities. This step is performed by production engineers and maintenance personnel. The early equipment management has several stages that include planning, design, fabrication, installation, test running, and commissioning. The goal of TPM is to maximize equipment effectiveness and pursue economic LCC.

12.5.4 Stabilization

- **Perfect TPM implementation and raise TPM levels**. This is the final step and at this stage, the organization perfects TPM implementation using continuous improvement tools such as Six Sigma and the Deming cycle.

12.5.5 Challenges and Success Factors of TPM Implementation

A proper implementation of TPM allows fundamental improvement within an organization by improving worker and equipment utilization. Improvement in equipment effectiveness and attitude of employees are key elements in the overall improvement within the organization (for more on TPM, see [3, 4]). However, TPM implementation is not an easy task and there are more companies that have failed in TPM implementation than those that have succeeded. TPM implementation usually encounters the following challenges:

- To secure and keep top management commitment.
- TPM implementation is a long-term project and takes around 3 years. It is a challenge to keep the momentum, resources, people motivation, and management commitment that long.
- TPM implementation requires organization culture change, and it is a challenge to change cultures. Autonomous maintenance where operators do maintenance tasks is hard for the organization especially those that have set division of labor polices and clear demarcation lines among departments.
- Lack of cooperation from operations/production may occur. As an example, in some cases production foremen are not willing to give up their people for TPM implementation training because they have a heavy work load. See John Auskamp [4] for more on how to make TPM everyone priority.
- Establishing incentives for the implementation process.

To ensure TPM implementation, the TPM project leadership must address the challenges stated above. The following are the success factors for TPM implementation:

- **Top management commitment and contributions**. This is achieved by managing expectations and demonstrating to top management the expected return on investment from the improvement on OEE. A pilot project for TPM implementation may be used to demonstrate TPM benefits.
- **Cultural transformation through training and reorganization**. This is achieved through training and education. Reorganization of maintenance and production departments is expected to help. Visits to organizations that won the TPM implementation prize are useful.

- **Employee involvement**. At the beginning, create a process for people to be involved because everybody wants to be of help and participate. In other words, involve everyone. The next step is to empower employees to make changes through a well-designed program.
- **Establishing planned and proactive maintenance polices**. Existing planned and scheduled maintenance should be evaluated as part of TPM implementation.
- **Training and education is important for TPM implementation**. It can be developed and organized by benchmarking with organizations who won the TPM prize or involving a consultant.
- **Communication**. In order to achieve ownership and the change needed to implement TPM, an environment of trust and goodwill must be created. It is necessary to create a flow of communication with operators, supervisors, and managers. When they perceive our sincere purpose of providing them with a better equipment and working environment, most of the barriers will be overcome.

For more on TPM implementation challenges and success factors, see Ahuja [1].

12.6 Summary

This chapter presented TPM. It provided a brief history of TPM followed by the goals and features of TPM. TPM has brought TQM techniques and tools to maintenance. Autonomous maintenance and equipment management are two unique features of TPM that are presented in detail in this chapter. An example how to compute OEE is provided in this chapter. A detailed account for TPM implementation following Nakajima approach in [5] is provided including challenges and success factors.

Exercises

1. Define TPM.
2. State the six big losses TPM strives to reduce and how would you measure each loss?
3. How would you develop an autonomous maintenance program?
4. Visit a plant near your area and select one of their critical equipment. Compute the overall effectiveness for this equipment.
5. State the steps of the PM methodology developed by JIPE.
6. What are the prerequisites for TPM implementation?
7. State the steps for TPM implementation.

8. Explain how the story of John Auskamp "How to Make TPM Everyone's Priority" helps in TPM implementation. Obtain the story and read it then answer.
9. Compare and contrast the steps for implementing TPM and TQM.
10. Suggest an effective organization for TPM implementation.
11. Design an education and training program for TPM implementation.
12. What is the major challenge in TPM implementation and suggest how to overcome it.
13. Select three companies that have won a TPM implementation prize and identify what are the common features among their maintenance systems.
14. How can RCM help in TPM implementation?

References

1. Ahuja PS (2009) Total productive maintenance. In: Ben Daya M, Duffuaa S, Raouf A, Knezevic J, Ait-Kadi D (eds) Handbook of maintenance management and engineering. Springer, London
2. Ben Daya M (2000) You may need RCM to enhance TPM implementation. J Qual Maintenance Eng 6(2):82–85
3. Hartman PE (1992) Successfully installing TPM in a non-Japanese plant. TPM Press Inc., Pittsburgh
4. John Auskamp in his story "How to make TPM everyone's priority". www.tpmconsulting.org/dwnld/article/tpmww/TPM%20article6.pdf
5. Nakajima S (1988) Introduction to TPM. Productivity Press, Cambridge

Chapter 13
Intelligent Maintenance

13.1 Introduction

The purpose of this chapter is to present e-maintenance and intelligent maintenance (*i*-maintenance) systems, a relatively new approach to maintain engineering and management that was coined by J. Lee around 2000. An intelligent maintenance system (IMS) is a system that utilizes the collected data from equipment and machinery in order to predict and prevent the potential failures [Lee et al. [6]. e-maintenance is the backbone of i-maintenance. Today, e-maintenance became a very common term in the maintenance literature and many companies such as Canon, American Petroleum Institute (API), British Petroleum (BP), and American Telephone & Telegraph (AT&T) is advertising on the Internet e-maintenance programs for their products. The following factors have contributed to the emergence of e-maintenance:

- The recognition of the role and impact of maintenance on business in a highly competitive market place.
- The need for high performance, near 100 % availability and zero product defects.
- The impact of maintenance on safety and the environment.
- The advancement in information and communication technologies (ICT).

The authors believe that condition-based maintenance is the root of e-maintenance and still shares several features that include predicting, preventing, and performing maintenance effectively and efficiently. ICT has enabled e-maintenance to integrate network of machines, and technology has enabled e-maintenance to share data effectively. This has added to the visibility of equipment health and performance. This visibility has made it easier to react and take maintenance tasks as needed at the right time. The elements of e-maintenance include the following:

S.O. Duffuaa and A. Raouf, *Planning and Control of Maintenance Systems*,
DOI 10.1007/978-3-319-19803-3_13

- **Base knowledge**: understanding the physical system to be maintained and its critical features and characteristics from which its performance and health can be predicted.
- **Data acquisition system**: an ICT system for monitoring the physical system to be maintained and collecting relevant data about equipment features and characteristics with the capability to share it over the intranet, extranet, and the Internet.
- **Mathematical and statistical models**: these are models to support maintenance decision making and that have the capability to estimate physical system reliability, remaining useful life, determining when maintenance action is needed, and plan and schedule the maintenance.
- **Performance reporting**: this is a system for reporting performance including e-maintenance, production, and other services.

The rest of this chapter is organized as follows: Sect. 13.2 presents predictive maintenance (PdM), and Sect. 13.3 covers the elements of e-maintenance. Section 13.4 outlines implementation of e-maintenance including requirements and barriers. Three case studies are presented in Sect. 13.5. Section 13.6 concludes this chapter.

13.2 Predictive Maintenance

PdM is a proactive type of maintenance that predicts failure or degradation state where maintenance action is necessary to minimize downtime and outage costs. PdM utilizes techniques such as vibration analysis, infrared thermography, tribology (oil analysis), noise and temperature to continuously monitor equipment degradation, and its evolution to predict failure. Several of these techniques were explained in Chap. 3. The maintenance action in PdM is based on the health condition of the equipment or the system and that is the reason it is known in the literature as condition-based maintenance (CBM). PdM has three main phases:

1. **Surveillance**—monitoring the equipment condition and collecting data that are relevant and can be used effectively to detect developing problems using various statistical and mathematical techniques.
2. **Diagnosis and prediction**—isolating the root cause of the problem and developing a maintenance action plan based on priority, equipment state of degradation, risks of failure, and its remaining useful life.

    1. **Correct and prevent**—Identify and perform corrective and preventive actions based on the outcome of step 2 above.

13.2.1 P–F Interval

The logic of PdM depends on the fact that most failures do not occur instantaneously, but in fact, equipment goes through a measurable process of degradation until they fail. Actually, there is a point in time where failure starts (not related to age and perhaps not detectable by existing technologies); however, there is a point P where humans can detect that failure has started and will take an interval till functional failure happens. This interval is known as the P–F interval, and Christer et al. [1] refer to it as delay time. Figure 13.1 illustrates the process of failure and shows the P–F interval.

Although many failure modes are not age-related, most of them give some sort of warning that they are in the process of occurring or about to occur. The frequency of predictive maintenance tasks has nothing to do with the frequency of failure and nothing to do with the criticality of the item. It is often possible to detect the fact that the failure is occurring during the final stages of deterioration, it may be possible to take action to prevent it from failing completely and/or to avoid the consequences. In practice, there are many ways of determining whether failures are in the process of occurring (e.g., hot spots showing deterioration of furnace refractories or electrical insulation, vibrations indicating imminent bearing failure, increasing level of contaminants in lubricating oil).

The P–F interval governs the frequency with which the predictive task must be done. The checking interval must be significantly less than the P–F interval if we wish to detect the potential failure before it becomes a functional failure. The P–F interval can be measured in any units relating to exposure to stress (running time, units of output, stop–start cycles, etc.), but it is most often measured in terms of

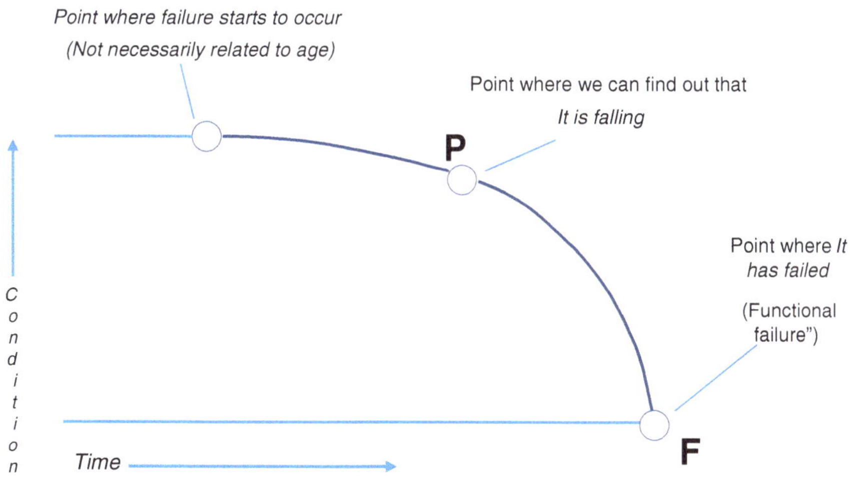

Fig. 13.1 P–F interval

elapsed time. The amount of time needed to respond to any potential failures which are discovered also influences condition-based task intervals.

The main idea behind PdM is to use the equipment or product degradation information to minimize downtime by ensuring that maintenance is taken at the right time, when it is needed and thereby avoiding under- and overmaintaining. Many techniques have been developed to help in features extraction, predicting failure or departure from acceptable performance. The techniques include signal processing technologies, neural networks, fuzzy logic, and decision trees.

13.2.2 Predictive Maintenance Implementation

The following factors are essential for a successful implementation of PdM. The factors are the main ingredient for the three phases of PdM mentioned in Sect. 13.2 above.

- Surveillance and monitoring:
  - Determine systems to be monitored based on knowledge of failures and their impact.
  - Determine data to be collected.
  - Determine expected frequency of data collection.
  - Develop systems for data acquisition.
  - Determine how to handle failures in the sensors and data acquisition system.
  - Determine how to handle imperfections in the received data, for example, a faulty sensor sending incorrect data.
  - Determine how to handle backup for the collected data.
  - Determine who should have access to the collected data.
- Diagnosis and prediction:
  - Determine models and techniques that represent assets/system behavior.
  - Determine acceptable behavior, anomaly, and failure.
  - Determine statistical and mathematical models that estimate and predict time and consequences of failure.
  - Determine optimal response (maintenance action) in case of any anomaly detection.
  - Determine how to address situations where there are multiple imminent failures that may occur.
- Correct and prevent:
  - Build the infrastructure needed for PdM,
  - Train maintenance staff in the requirements of PdM,
  - Conduct change management process required for PdM,

- A pilot implementation may help,
- Implement maintenance actions, and
- Evaluate effectiveness and improve.

13.3 Elements of Integrated e-maintenance

Intelligent prognostics is defined by Lee et al. [6] as a systematic approach that can continuously track health degradation and extrapolate temporal behavior of health indicators to predict risks of unacceptable behavior over time as well as pinpoint exactly which components of machines are likely to fail. Intelligent prognostics coupled with information flow in which maintenance actions are synchronized with the operation of the system as well as the maintenance resources has enabled maintenance experts to move from PdM to e-maintenance. The synchronization of maintenance actions and information flow infrastructure enhances visibility and enables autonomous triggering of services and ordering of parts. Such an integrated system with a high visibility is expected to result in near 100 % availability.

The key elements of e-maintenance are the following:

- Micro-electromechanical and wireless sensors,
- Web-based and semantic maintenance technologies and tools,
- Mobile devices,
- Asset self-identification technologies, such as RFID, and
- Computing facilities equipped with statistical and mathematical models for data processing and analysis.

e-maintenance employs the above technologies, know-how, and software to integrate maintenance stakeholders, tools, processes, and data. This integration makes equipment conditions and decisions to be made visible to maintain technical staff that is enabled to take cost-effective maintenance actions when needed and at the right time. The elements of e-maintenance mentioned above provide services such as maintenance documentation, predictive health monitoring and maintenance planning services, performance assessment integration, as well as training and knowledge management at the machine level under dynamic conditions. Their range includes services to deliver anywhere, any time, and to anyone authorized to have access.

13.4 Implementation of e-maintenance

The implementation of e-maintenance will draw from Zachman four levels framework [8]. The framework starts from top to bottom. The levels are as follows:

- e-maintenance strategic vision that must be aligned with the business strategy and goals. The vision supports the e-maintenance scope.

- e-maintenance business processes: specify the processes necessary to support the deployment of e-maintenance including training and change management.
- e-maintenance organization: define the organization for e-maintenance and the changes to deploy functions to implement e-maintenance.
- e-maintenance IT infrastructure, which supports the organization and implements e-maintenance processes. This level materializes the information technology means required (ICT requirements) for running applications and for enabling communication between these applications according to their distribution on site.

13.4.1 Requirements for Implementation

e-maintenance can be viewed as a new strategy for maintenance or a major project for the organization to undertake. Whether it is viewed as a strategy or a major project, it is expected to transform the organization. The implementation of e-maintenance has special requirements to avoid failure. The requirements include the following:

- It requires management commitment and support because the implementation of e-maintenance requires financial support over a long period of time to build needed infrastructure.
- The level of maturity and sophistication of the current maintenance strategies. The organization must have implemented preventive and predictive maintenance.
- Having the right organization for e-maintenance or at least part of it.
- Building the right information and communication technologies infrastructure necessary for data collection, analysis, and sharing with authorized personnel.
- Training of staff to bring them to the level to use the new infrastructure and to be effective decision makers. For more on e-maintenance, refer to [2, 6, 7].

13.4.2 Challenges of Implementation

The challenges are related in a sense to the requirements; however, in this part, the major challenges are listed. The major challenges are related to people and processes and include the following:

- Human resource restructuring, and training [9]. Each maintenance actor (technician, engineer, or leader) has to become capable of pacing with the speed of information flow and understanding the overall e-maintenance structure.

- Risk management in e-maintenance activities involves a trade-off between protection on the one hand and functionality, performance, and ease of use on the other [10];
- The development of a stable distributed computing, optimization, and synchronization system for dynamic decision making [5];
- The security and reliability concern arising from transactions over the Internet [3]; and
- Manage to predict failures and disturbances, and to estimate the remaining lifetime of components, mechanical systems, and integrated systems [4].

13.5 Case Studies

The details of several successful case studies for e-maintenance and intelligent maintenance decision-making tools have been reported in Lee et al. [6]. The cases are briefly highlighted below and they include the following:

13.5.1 Roller Bearing Performance Predication

Using vibration wave form coupled with wavelet filter designed to denoise the raw signal has increased the probability of detecting degradation. This enhanced the prediction of bearing actual life.

13.5.2 Industrial Network Fault Detection

A large number of industrial network failures are attributed to loose or degrading terminating resistors. This case from an experiment conducted at the University of Michigan using a controller area network (CAN) test setup from an industrial sponsor. Normal signals were acquired using a properly functioning terminating resistor and then faulty signals were created by removing one terminating resistor from the network. The CAN signal is a logical differential signal, with a high- and a low-voltage component. One way to detect a faulty resistor in a network that emits this kind of signal is to measure the overshoot and the signal-to-noise ratio of the logical 0s and 1s in the signal. Since the changes happen very quickly, the signal must be sampled at a very high sampling rate. Here, we used a sampling rate of 20 MHz. for manipulation of the network data, the team conducting the experiment did not use the frequency-based methods for feature extraction, instead the overshoot and signal-to-noise ratio are used as expert extracted features. These expert extracted features are taken from 10 normal and 10 faulty signals for training. Using these signals, it is possible to use logistic regression to classify the 200 measured signals as normal or faulty. The data cannot be successfully classified using statistical pattern recognition.

13.5.3 Maintenance Scheduling Using Predictive Information

Discrete-event simulation and heuristic optimization are utilized for scheduling of maintenance operations that are the least intrusive on the normal production operation in a manufacturing system, as suggested in [11]. The cost-effectiveness of many maintenance schedules were evaluated based on predicted probabilities of equipment failures over time, obtained from predictive condition-based algorithms. In this case, the impact of equipment failures and maintenance operations was assessed through discrete-event simulations, and a GA-based search algorithm was used to search for maintenance schedules with highest corresponding average cost-benefits. The following is a short description of an example given in [11] where advantages of the newly proposed maintenance scheduling method over more traditional methods are demonstrated.

Four types of maintenance strategies have been simulated and compared with the maintenance schedule obtained using the method from [11]. The four maintenance strategies considered in this example were as follows:

- *Corrective maintenance strategy*, which uses the simple first-come–first-serve scheme in which the maintenance is performed whenever there is a machine failure and there is a maintenance person available. If any of the two conditions is not satisfied, the machine will remain in a failed mode, not producing anything. This maintenance strategy will be referred to as “Strategy A”;
- *Scheduled maintenance strategy*, in which maintenance is performed in regular time intervals. This strategy will be referred to as “Strategy B”;
- *Condition-based maintenance*, in which maintenance crews possess information about the current condition of the equipment. Thus, instead of waiting for machine failure, it is assumed that user-defined thresholds are set on the degradation level of any given machine to trigger maintenance operations. This strategy will be referred to as “Strategy C”;
- *Predictive maintenance strategy* based on the maintenance scheduling methods, using the current and predicted equipment conditions and taking into account both production benefits and maintenance expenses. This strategy will be referred to as “Strategy D.”

The simulation study has shown that strategy D is superior in all test cases.

13.6 Summary

This chapter presented e-maintenance and intelligent maintenance systems. Two main factors lead to the emergence of e-maintenance. The first one is the recognition of the role and impact of maintenance on the enterprise business. The second is the advancement of the information and communication technologies (ICT) including computing facilities and mobile devices. The ICT leads to visibility of equipment health, spare parts inventories, and other operations related data. The visibility has provided stakeholders (actors) with the ability to respond at the right time to prevent failure and reduce downtime to near zero. This chapter includes PdM in brief, elements of e-maintenance requirements, and challenges for e-maintenance implementation. In addition, three cases are outlined to demonstrate some success stories.

Exercises

1. What is e-maintenance?
2. Define the term intelligent prognostics.
3. What factors led to the development of e-maintenance?
4. How does ICT support e-maintenance?
5. List the steps for implementing PdM.
6. What is the P–F interval and how does it relate to the concept of delay time introduced by Christer?
7. What are the requirements to initiate an e-maintenance program?
8. What are the challenges for implementing e-maintenance?
9. Review reference number 6 in the list above and extract the main elements of e-maintenance and compare it to the one listed in Sect. 13.3.
10. Select an organization/plant in your area and assess if it is ready to initiate e-maintenance.

References

1. Christer AH, Walker WM (1984) Delay time models for industrial inspection maintenance problem. J Oper Res Soc 35(5):401–406
2. Djurdjanovic D, Lee J, Ni J (2003) Watchdog agent: an infotronics-based prognostics approach for product performance degradation assessment and prediction, special issue on intelligent maintenance systems. Eng Inf J 17(3–4):107–189 (formerly AI in Engineering)
3. Hamel W (2000) E-maintenance robotics in hazardous environments. In: Proceedings of the 2000 IEEE/RSJ international conference on intelligent robots and systems, Takamatsu, Japan
4. Holmberg K, Helle A, Halme J (2005) Prognostics for industrial machinery availability. In: POHTO 2005 international seminar on maintenance, condition monitoring and diagnostics, Oulu, Finland

5. Koc- M, Ni J, Lee J, Bandyopadhyay P (2003) Introduction of e-manufacturing. In: Proceedings of the 31st North American manufacturing research conference (NAMRC), Hamilton, Canada
6. Lee J, Ni J, Dragan D, Hai Q, Haitao L (2006) Intelligent prognostics and e-maintenance. Comput Ind 57:476–489
7. Liyanage JP, Jay L, Emmanouilidis C, Ni J (2009) Integrated e-maintenance and intelligent maintenance systems. In: Ben-Day M, Duffuaa S, Raouf A, Knezevic J, Ait-Kadi D (eds) Handbook of maintenance management and engineering. Springer, Berlin, pp 499–539
8. Mullera A, Crespo Marquezb A, Lunga B (2008) On the concept of e-maintenance: review and current research. Reliab Eng Syst Safety 93:1165–1187
9. Ong MH, Lee SM, West AA, Harrison R (2004) Evaluating the use of multimedia tool in remote maintenance of production machinery in the automotive sector. In: IEEE conference on robotics, automation and mechatronics, vol. 2, Singapore, pp 724–728
10. Tsang A (2002) Strategic dimensions of maintenance management. J Qual Maint Eng 8(1):7–39
11. Yang Z, Djurdjanovic R, Ni J, Lee J Maintenance scheduling in production systems based on predicted machine degradation. IEEE Trans Autom Sci Eng. Paper No. V2004-067 (submitted for publication)

Chapter 14
Maintenance System Performance, Productivity, and Continuous Improvement

14.1 Introduction

Productivity is defined as output per unit of input. In a maintenance system, the output is sustainable productive capacity of the equipment being maintained, and the input is the resources required to sustain that capacity. As one ratio of productivity for maintenance is not practical, a separate list of input measures, output measures, and measures within the maintenance system is required.

Continuous improvementof the maintenance systems may be achieved using three approaches. The first approach is to use performance measures, and the second approach is to conduct periodic auditing of the maintenance system. The third approach is to combine both approaches. In the first approach, performance measures for the maintenance system are defined and computed on periodic basis. Targets are set, and based on the system performance and the targets, improvement programs are developed and implemented. In the second approach, the regular audits are used to identify areas to improve the system. Productivity and continuous improvement of the maintenance systems is the subject of this chapter.

The rest of the chapter is organized as follows: Section 14.2 describes a set of input, output, and within the system measures together with the balanced scorecard. Section 14.3 presents maintenance indices for administration, effectiveness, and cost followed by metric comparisons in Sect. 14.4. Section 14.5 focuses on maintenance auditing, and Sect. 14.6 presents root cause analysis for developing corrective and preventive actions that are expected to result in continuous improvement. Section 14.7 describes a case study for maintenance auditing followed by benchmarking in Sect. 14.8. Section 14.9 highlights techniques for improving maintenance performance and costing. The chapter is concluded by a summary and a set of exercises.

S.O. Duffuaa and A. Raouf, *Planning and Control of Maintenance Systems*,
DOI 10.1007/978-3-319-19803-3_14

14.2 Maintenance Performance Measures

Performance measures (PM) are useful for improving performance and may be used for continuous improvement. The PM are divided according to the purpose of the analysis. In this section, the measures are divided into input, output, and within the system measures. The next sections present the three types of measures followed by the balanced scorecard approach for developing performance measures.

14.2.1 Input Measures

Some of the important factors Measureswhich are directly related to the cost of performing maintenance are stated below:

- **Labor**: This includes all costs associated with the trades, apprentices, semi-skilled support labor, and custodial and grounds staff directly employed by the maintenance department. These costs also include overtime, training, benefits, and various mandated or regulatory costs.
- **Materials**: This covers purchased parts, spares, supplies, stationery, safety protective wear, shop supplies, and chemicals used directly for repair and maintenance activities. This can also include direct material overhead applied to the maintenance stores issue price, such as transport, storage, handling, shipping, and internal parts delivery.
- **Contracts**: This covers the cost of specific maintenance tasks, projects, or work orders contracted on a time or project basis, such as facility maintenance, boiler overhaul, or machine refurbishment, which are all included under this head. Each work category is characterized by a specific contract. This is in addition to the cost of temporary or permanent contracted personnel supporting the directly employed personnel.
- **Shop services**: In larger facilities and plants, there is usually a central shop service which provides specialized or trade-specific services. These often charge a calculated hourly rate to specific work orders or projects as directed by the requesting area. This hourly rate generally incorporates all the shop overhead such as supervision and building and equipment operating costs.
- **Equipment rentals**: This covers the cost of all mobile and stationary equipment rentals such as cranes, flatbeds, backhoes, and hydraulic rams.
- **Tool crib**: This covers the cost of specialized hand tools and specialty tools such as pneumatic wrenches, chainfalls, hand grinders, and acetylene torches, if not included in materials (stores) or in shop services.
- **Maintenance overhead**: This may include all levels of maintenance management and supervision as well as maintenance and reliability engineering, planners, schedulers, material coordinators, clerks, and data input and computerized maintenance management systems' support. It may also include any shops' costs not included in shop services.

- **Company or plant overhead**: This normally includes a proportion of all company management and executive overhead costs, including depreciation.

Along with the costs of the above categories of input measures, there may be specific non-cost measures for each. For example, when looking at materials, several statistics are useful for stores inventory control:

- Inventory investment or value by class or category.
- Inventory turns, defined as the ratio of total annual issues divided by the on-hand quantity at year end.
- Service level, defined as the percentage of stores orders filled as requested.
- Stock-outs, defined as the number of orders unable to be filled as requested from the stores over a specified time period.
- Inventory obsolescence, defined as the percentage of stock keeping units (SKUs) which have had no issue within a 24-month period.

14.2.2 Output Measures

Output measures describe why maintenance management exists and include the following measures:

- **Availability**: It is a measure of uptime, or alternatively a measure of the duration of downtime, defined as (scheduled time minus all delays)/scheduled time.
- **Reliability and mean time between failures** (**MTBF**): It is a measure of the frequency of failure, defined as running time/number of failures.
- **Mean time to repair** (**MTTR**): It is a measure of the duration of repair time, defined as repair downtime/number of failures. Maintainability is the probability of performing the repair in a given time or at MTTR.
- **Process rate**: It is a measure of the cycle time of the equipment in the process, defined as the ideal cycle time/actual cycle time or alternatively the actual throughput rate/ideal throughput rate. With this measure, it is often easier to define the "ideal" as the statistical upper control limit for the particular process.
- **Quality rate**: It is a measure of the process or equipment precision, defined as (total throughput minus net rejects)/total throughput. The net rejects include the loss for recycled, rejected, scrapped, or downgraded product.
- **Overall equipment effectiveness** (OEE): It is the product of availability, process rate, and quality rate; it is a cross-functional measure, as several departments or functions can have an impact on results.

The next section demonstrates the estimation of these output measures and their impact on the productive capacity. Consider a piece of production equipment with the operating and maintenance history described in Fig. 14.1 over a one-month period.

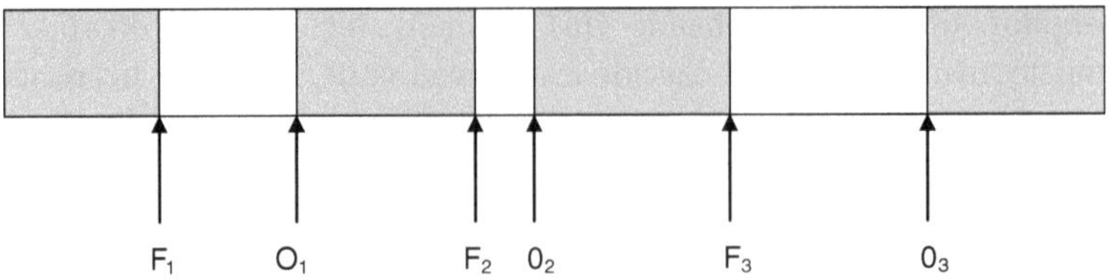

Fig. 14.1 Maintenance measures

In this scenario, the production schedule calls for continuous operation throughout the month; however, equipment is not available to produce throughout that period.

Availability is a measure of the duration of scheduled operation that is achieved, or conversely, a measure of the duration of failure or downtime is given by:

$$A = \frac{S - d}{S} \times 100\,\% \tag{14.1}$$

where

A = availability
S = scheduled production time
d = downtime in days

$$A = \frac{31 - 6}{31} \times 100\,\% = 80.6\,\%$$

In this example:

Id = 100 % − A = 100 − 80.6 = 19.4 %
Id = percentage of downtime

Most operating environments are sensitive to the frequency of failures. In those cases, because of associated production and quality issues, one shutdown for two days is preferable to four shutdowns, each for a half a day. Therefore, a measure of the frequency of failure is required:

MTBF = Mean time between failures

$$\text{MTBF} = \frac{S - d}{f} \tag{14.2}$$

f = Number of failures

$$\text{MTBF} = \frac{31 - 6}{3} = 8.33\,\text{days}$$

Once a piece of equipment or component has failed, it is normally critical to minimize the repair time. The measure of repair time is MTTR. Maintainability is the probability that equipment will be repaired within MTTR.

$$\text{MTTR} = \frac{\mathrm{d}f}{f} \tag{14.3}$$

MTTR = Mean time to repair
df = downtime delays from failures

For the above example:

$$\text{MTTR} = \frac{6}{3}$$

Clearly, in this example, the equipment would be considered unreliable (MTBF = 8.3 days) and MTTR = 2 days, although the availability at 80.6 % would not be considered particularly unusual.

Further measures must be considered to fully describe the productive capacity and capability of an operating system. Consider the following scenario:

A cement kiln has a requirement to operate 300 days per year, producing at a rate of 1200 tons per 24 h day. Over the years, its performance has been given as follows.

- Scheduled operation time is 300 days,
- Number of breakdowns are 6,
- Delays for breakdowns accounted for 180 h,
- Preventive maintenance time is 80 h,
- Shutdown due to lack of feed lasted for 120 h,
- Process slowdowns are for 60 h at 80 % rate,
- Wet feed slowdowns are 40 h at 60 % rate,
- Total rejected production is 2000 tons, and
- Second-quality production is 3000 tons at 50 % value.

When reviewing production capacity of the plant, the scheduled time relative to calendar time is a measure of its utilization.

$$\begin{aligned} \text{Utilization} &= \frac{\text{Scheduled Time}}{\text{Calendar Time}} \times 100 \\ &= \frac{300}{365} = 82.2\,\% \end{aligned} \tag{14.4}$$

The measure of availability for this plant is calculated as follows:

$$\begin{aligned}\text{Availability} &= \frac{300 - (180 + 80 + 120)/24}{300} \times 100 \\ &= 94.7\,\%\end{aligned}$$

As production and maintenance systems have become more complex with a higher requirement for integration, it has become useful to measure all delays and losses with a consistent approach, regardless of whether the operations or maintenance department has had the traditional responsibility for its correction.

We see in the example that production capacity has been diminished because the kiln was run at slower speeds for a total of 100 h, resulting from both process and feed problems. Often, equipment cannot be run to its required rate because of inadequate operating procedures, maintenance activities, or design problems.

The process rate in this example is

$$\begin{aligned}\text{Process Rate} = \text{PR} &= \frac{\text{Ideal Cycle Time}}{\text{Actual Cycle Time}} \times 100\,\% \\ \text{or, PR} &= \frac{\text{Actual Operating Throughput Rate}}{\text{Ideal Operating Throughput Rate}} \times 100 \\ &= \frac{(300 - 380/24) \times 1200 - (60/24 \times 0.2 \times 1200) - (40/24 \times 0.4 \times 1200)}{(300 - 380/24) \times 1200} \times 100\,\% \\ &= \frac{284.167 \times 1200 - 2.5 \times 0.2 \times 1200 - 1.67 \times 0.4 \times 1200}{284.2 \times 1200} \times 100\,\% \\ &= \frac{341{,}000 - 600 - 80}{341{,}040} \times 100\,\% \\ &= \frac{339{,}600}{34{,}100} \times 100\,\% \\ &= 99.6\,\%\end{aligned} \tag{14.5}$$

Finally, although products were produced, all of them cannot be sold, again reducing the productive capability.

$$\begin{aligned}\text{Quality Rate} = \text{QR} &= \frac{\text{Total Produced} - \text{Reject Downgrade}}{\text{Total Produced}} \times 100\,\% \\ &= \frac{339{,}600 - (2000 + 3000)}{339{,}600} \times 100\,\% \\ &= \frac{334{,}600}{339{,}600} \times 100\,\% \\ &= 95.5\,\%\end{aligned} \tag{14.6}$$

A true measure of the OEE of the kiln is:

$$\begin{aligned} \text{OEE} &= A \times \text{PR} \times \text{QR} \\ &= 0.9470 \times 0.995 \times 0.985 \times 100\,\% \\ &= 92.9\,\% \end{aligned} \tag{14.7}$$

The production capacity of the equipment must reflect the utilization of the kiln and can be obtained as follows:

$$\begin{aligned} \text{Productivity Capacity} &= \text{OEE} \times \text{Utilization} \\ &= 0.929 \times 0.822 \times 100\,\% \\ &= 76.4\,\% \end{aligned} \tag{14.8}$$

Therefore, the kiln has an additional capacity of 23.6 %, which may be achieved by managing the plant production schedule, the delays, its speed, and the product quality.

14.2.3 *Measures Within the System*

Although productivity looks at only inputs and outputs, impacting either of these requires a clear understanding of the system itself and how the productivity measures relate to the system measures.

- Work distribution: This is the percentage of labor hours spent on the various categories of work such as planned, unplanned, urgent, minor repair, standing order, indirect (e.g., meetings, training), preventive, predictive, planned corrective, major overhaul, or shutdown work.
- Delays: This is the time spent by trades waiting for instructions, for parts, for other trades, travel time, personal breaks, start and quit times.
- Compliance: These are measures to track compliance with various plans and schedules and include PM coverage, PM schedule compliance, corrective planned work schedule compliance, work requests generated from PM routines, and shutdown schedule compliance.
- Backlogs: This is the amount of work planned but not yet scheduled or completed, usually calculated by trade (mechanical, electrical), by plant or shop area, by shift, or by specific crew.
- Work order status: This is a measure of the number of work orders or work requests in each state of completion: received by the maintenance department, approved, planned, waiting on materials, scheduled, assigned, in progress, or completed.
- Failure analysis: This is an activity area that tracks improvement initiatives, such as the number of root cause analyses of breakdowns undertaken and completed, the number of PM routines developed, warranty issues, and the like.

The above sections provide an overall definition of the important factors that should be considered in evaluating the effectiveness and productivity of a maintenance system. The next section gives a set of useful indices that measure the effectiveness of these factors.

14.3 Maintenance Indices

There are several indices that measure the performance of maintenance. They are useful in reports preparation and provide a reasonable quantification of the performance of some key areas. These indices should be integrated with the computerized maintenance management system, and the system should provide them automatically when requested. Broadly, these indices can be classified into two categories. In the first category are economic (cost) indicators that allow the follow-up of the evolution of internal results and certain comparison between maintenance at different plants. The second are technical indicators that give the maintenance manager the means of following up the technical performance of the installations. Niebel [4] assigned these indices to three classes that are related and reflect the maintenance objectives. These classes are maintenance administration, maintenance effectiveness, and maintenance costs. The following list of indices must be customized for each organization prior to use.

14.3.1 Maintenance Administration

1. Subcontracted hours per month.

$$\% = \frac{\text{Total subcontracted hours worked}}{\text{Total hours worked}} \times 100$$

2. Overtime hours per month

$$\% = \frac{\text{Total overtime hours worked}}{\text{Total hours worked}} \times 100$$

3. Craftsmen activity level

$$\% = \frac{\text{Standard hours earned}}{\text{Total check time}} \times 100$$

4. Current backlog (in crew-weeks)

$$\text{Crew weeks} = \frac{\text{Work scheduled ready to release (in man-hours)}}{\text{One crew week (in man-hours)}}$$

5. Total backlog (in crew-weeks)

$$\text{Crew weeks} = \frac{\text{Total man-hours of work awaiting execution}}{\text{One crew week (in man-hours)}}$$

6. Craftsmen productivity per month

$$\% = \frac{\text{Total hours worked}}{\text{Standard hours}} \times 100$$

7. Craftsmen utilization

$$\% = \frac{\text{Hours spent on product work}}{\text{Total hours scheduled for work}} \times 100$$

8. Composite Productivity Index (CPI) for craftsmen

$$\text{CPI} = \text{Productivity} * \text{Utilization}$$

9. Work orders planned and scheduled daily

$$\% = \frac{\text{Work orders planned and scheduled}}{\text{Total work orders executed}} \times 100$$

10. Scheduled hours versus hours worked as scheduled

$$\% = \frac{\text{Work orders planned and scheduled}}{\text{Total work orders executed}} \times 100$$

11. Scheduled hours versus hours worked

$$\% = \frac{\text{Hours worked as scheduled}}{\text{Total hours scheduled}}$$

12. Preventive and predictive maintenance conducted as scheduled

$$\% = \frac{\text{Hours scheduled}}{\text{Total hours worked}} \times 100$$

13. Predictive and preventive maintenance coverage

$$\% = \frac{\text{Total man-hours of preventive and predictive maintenance executed}}{\text{Total man-hours worked}}$$

14.3.2 Maintenance Effectiveness

1. Overall equipment effectiveness (OEE)

$$\mathrm{OEE} = A \times S \times Q$$

A = Availability indicator

S = Speed indicator
Q = Quality indicator

2. $A = \dfrac{\text{Planned production time} - \text{Unplanned downtime}}{\text{Planned production time}}$

3. $S = \dfrac{\text{Actual amount of production}}{\text{Planned amount of production}}$

4. $Q = \dfrac{\text{Actual amount of production} - \text{Unaccepted amount}}{\text{Actual amount}}$

5. Percentage of gross operating hours

$$\% = \frac{\text{Number of gross operating hours}}{\text{Number of gross operating hours} + \text{downtime for maintenance}} \times 100$$

6. Number of failures in the system (NFS)

$$\mathrm{NFS} = \frac{\text{Number of production stops}}{\text{Number of gross operating hours}}$$

7. Equipment downtime caused by breakdown

$$\mathrm{NFS} = \frac{\text{Downtime caused by breakdown}}{\text{Total downtime}} \times 100$$

8. Emergency man-hours

$$\% = \frac{\text{Man-hours spent on emergency jobs}}{\text{Total direct maintenance hours worked}} \times 100$$

9. Emergency and all other unscheduled man-hours

$$\% = \frac{\text{Man-hours of emergency and unscheduled jobs}}{\text{Total maintenance man-hours worked}} \times 100$$

10. Evaluation of predictive and preventive maintenance

$$\% = \frac{\text{Job resulting from inspections}}{\text{Inspections completed}} \times 100$$

14.3.3 Maintenance Costs

1. Cost of maintenance to add value of production

$$\% = \frac{\text{Direct cost of maintenance}}{\text{Added value of production}}$$

   The direct cost of maintenance consists of the cost of manpower, cost of materials (spare parts, lubricants, etc.), and cost of subcontracted work and overloads. The added value of production is the cost of production less than the cost of material.

2. Maintenance cost per unit of production

$$\text{Cost per unit} = \frac{\text{Total maintenance}}{\text{Total units produced}}$$

3. Manpower component in the maintenance cost

$$\% = \frac{\text{Total maintenance manpower}}{\text{Total direct maintenance cost}} \times 100$$

4. Cost of subcontracted maintenance

$$\% = \frac{\text{Cost of subcontracting (manpower)}}{\text{Direct cost of maintenance}} \times 100$$

5. Ratio of labor cost to material costs of maintenance

$$\text{Ratio} = \frac{\text{Total maintenance labor costs}}{\text{Total maintenance material costs}}$$

6. Cost of maintenance hour

$$\$ = \frac{\text{Total cost of maintenance}}{\text{Total man-hours worked}}$$

7. Percent of supervision costs of total maintenance cost

$$\% = \frac{\text{Total cost of supervision}}{\text{Total cost of maintenance}} \times 100$$

8. Progress in cost reduction effects

$$\text{Index} = \frac{\%\,\text{maintenance man-hours spent on scheduled jobs}}{\text{Maintenance cost/Unit of production}}$$

9. Preventive maintenance (PM) cost as related to breakdown maintenance

$$\% = \frac{\text{Total PM costs (including production losses)}}{\text{Total breakdown costs}} \times 100$$

10. Inventory turnover rate per year

$$\text{Rate} = \frac{\text{Annual consumption costs}}{\text{Average investment inventory}}$$

11. Cost of spare parts and material to maintenance cost

$$\% = \frac{\text{Total store issues and purchases}}{\text{Total direct maintenance costs}} \times 100$$

12. Ratio of stock value to production equipment value

$$\text{Ratio} = \frac{\text{Average stock value}}{\text{Replacement value of production equipment}}$$

14.4 Metric Comparisons

Organizations develop performance measures based on the specific objectives sought at particular periods in corporate history. If, for example, a mining operation has a specific problem with mechanical failures in its haul truck fleet, a measure of availability is defined which reflects this concern. As a result, availability may be defined as calendar time minus delays for mechanical failures, divided by calendar time. This metric is useful to this mining operation, but not readily comparable to historical measures based on a definition that may have included all maintenance delays, nor to other mining companies or industry at large, who may use scheduled time rather than calendar time.

A measure should be defined so that it is comparable to external data, is clear and easily understood, uses readily available data, and has a broad consensus that it accurately measures the characteristic under consideration.

Earlier, the input, output, and system measures were discussed. Many of these measures can be compared among similar organizations, provided a rigorous definition is applied. Some of the metrics frequently compared in maintenance involve longer term or strategic ratios of costs, manpower, and materials:

- Total maintenance costs per the replacement value of the assets maintained,
- Spare parts and materials in inventory cost per the replacement value of the assets maintained,
- Spare parts and materials in inventory cost per number of maintenance tradesmen,

- Maintenance tradesmen per total direct operating labor employees,
- Maintenance tradesman per first-line supervisor,
- Number of organization levels directly involved in maintenance,
- Maintenance tradesmen per support staff (clerks, planners, engineers, schedulers, etc.), and
- Training cost/maintenance tradesmen.

Comparisons must always be tempered with an understanding of the differences in operating environment, such as political and regulatory environment, country and company culture, geographical remoteness, industry sector, asset age, and use of technology.

14.5 Maintenance Auditing

In this section, a step-by-step continuous improvement program for the maintenance systems is presented. The goal of the program is to achieve and establish a productive maintenance system, based on the above factors. The starting point in the design of an improvement program is to assess the current status of the system. The assessment is accomplished by an audit scheme, which consists of two steps. The first step is the scoring of the essential factors in the maintenance system. The second step is to obtain an audit score. In order to obtain an audit score [maintenance productivity indexed (MPI)], the weight of each factor or its contribution to the maintenance system must be determined. Analytic hierarchy process (AHP) provides a useful methodology for determining the weight of each factor. Combining steps 1 and 2, an audit score or MPI can be obtained. After auditing the system, the third step of the improvement program is to determine what the system's major unproductive factors are, and ABC analysis is used for that purpose. Following the identification of the system's crucial unproductive factors, a cause and effect analysis must be conducted to identify possible corrective action. After identifying corrective actions and their implementation, the program may be repeated to re-evaluate the system. Statistical tests of hypotheses are recommended to test for significant improvement. The complete plan is given in Fig. 14.2. The aim of this chapter is to present in detail the steps of the plan and demonstrate its utilization by a case study.

14.5.1 Factors in the Audit Scoring Scheme

In this section, the most important factors that influence maintenance productivity are outlined. These factors constitute the basis for the audit program. The importance and the impact of each factor on maintenance system productivity is briefly outlined. In the scoring scheme, each factor is scored out of 10, 10 being the

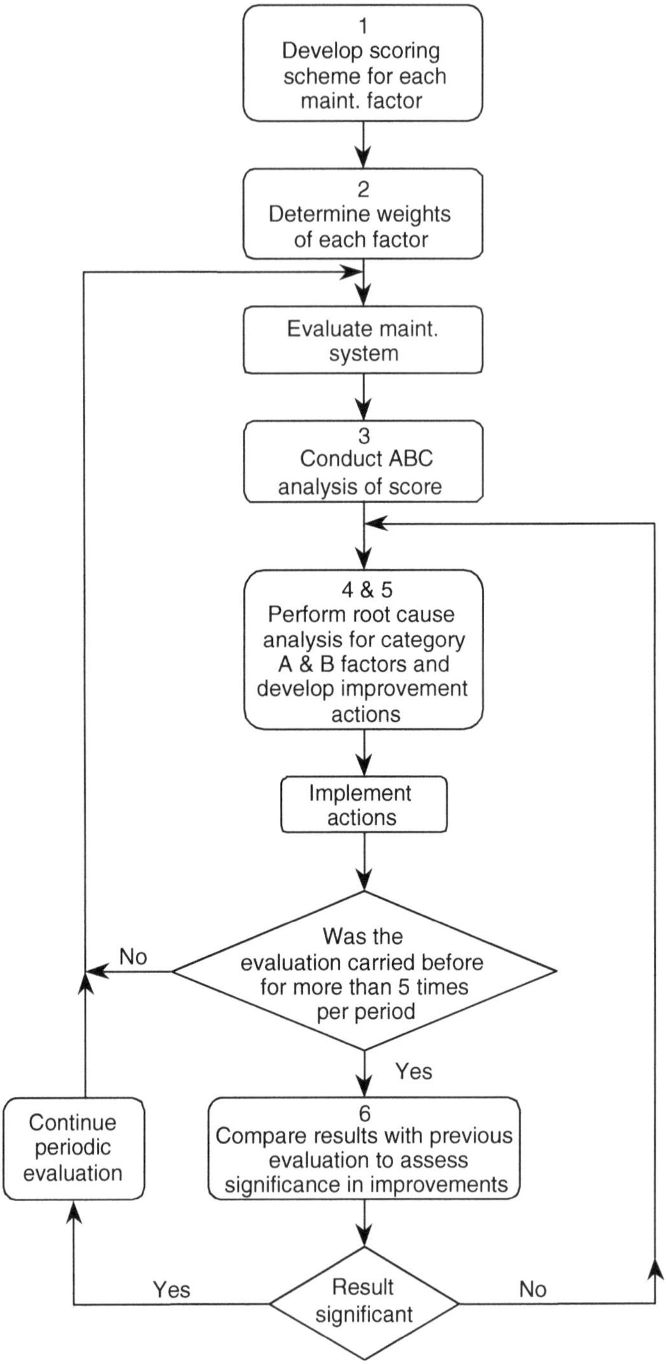

Fig. 14.2 Continuous improvement plan for the maintenance systems

highest. The scoring is done by examining the state of each factor in a maintenance system. A common approach is to develop a set of questions about each factor and then give a score for the factor from the answers to these questions. An approach similar to the one in several of Refs. [1, 3, 6, 7] can be employed to obtain the score for each factor. The major factors that are used in an audit scheme are discussed below.

Organization staffing and policy

Organizations are designed to facilitate the execution of maintenance plans. They spell out the responsibilities, chain of command, and span of control. Information flow and the ability to carry out specified plans are heavily affected by the organizational structure. The importance of this factor stems from the need to have a well-designed organizational structure and an effective span of control. The need for proper job descriptions and an organization chart will be recognized during the process of scoring this factor.

Labor productivity

Labor productivity is defined as the ratio of standard hours of work to the actual hours. This factor focuses on craft productivity. In the process of scoring this factor, management can identify crafts with low productivity and establish reasons for the low productivity. Training or higher skill level craftsmen may be required to improve the productivity of the maintenance systems.

Management training

This factor assesses the need for management training, especially in job standards, planning tools, and productivity improvement techniques.

Planner training

Proper planning and scheduling of maintenance jobs have a great impact on the productivity of a maintenance system. The planner plays an important role in the planning and scheduling of maintenance jobs and must be adequately trained.

Craft training

Training is an important supporting function for the maintenance system. This factor has a direct impact on factor number one. A well-defined training program must be established for each craft. The program should be updated on a yearly basis to reflect organization needs.

Motivation

Maintenance system productivity is highly dependent on manpower. The productivity and quality of an individual's performance are significantly affected by his morale. Therefore, high morale and motivation are important for improving productivity. When examining this factor, issues such as turnover rate and job security should be investigated.

Management control and budget

The need for control reports for budget control and equipment performance is addressed in this factor. Equipment downtime and backlog reports are good indicators of the effectiveness of a maintenance system.

Work order, planning, and scheduling

The work order system is the heart of any maintenance control system, and it is a necessary tool for effective planning and scheduling. This factor emphasizes the need for written work orders and the proper planning and scheduling of jobs. Planning and scheduling is the backbone of any maintenance system. In addition, this factor addresses the quality of maintenance jobs.

Facilities

This factor addresses the effect of proper layout of maintenance shops and good housekeeping on the productivity of a maintenance system. It also addresses the availability of necessary tools and equipment.

Stores, material, and tool control

This factor involves the inventory and tool control procedures. It emphasizes the need for an up-to-date inventory system and clear policies and procedures for tool management. Spare parts availability and bench stock management are essential for a productive maintenance system.

Preventive maintenance and equipment history

Preventive maintenance is an important element of any maintenance strategy. It is the action taken to prevent failure and provide means of controlling downtime and maintenance planning and scheduling. Historical data on equipment failure are the backbone of any statistically based preventive maintenance. This factor reflects heavily on the maintenance capabilities to prevent unexpected failures. Scoring this factor will identify the need for improving the preventive maintenance program.

Engineering and condition monitoring

This factor emphasizes the need for using diagnostic routines and establishing a condition-based maintenance (CBM) program which is essential for predictive maintenance.

Work measurement and incentives

This factor deals with developing standard times for standard jobs. Standard times are essential for planning and controlling maintenance work. Also, they can be used for assessing productivity. Scoring this factor will identify the need for developing maintenance standard times, revising, and/or updating existing ones.

Information system

An information system is a tool for proper management and control. It must be designed so that it satisfies maintenance management requirements. It has a significant impact on maintenance systems. It should have all the necessary subsystems that provide equipment, workload, and spares control, in addition to a timely reporting system.

A high score for the above fourteen factors is necessary for a productive maintenance system. A deficient factor or one with a low score will lower the effectiveness of the maintenance system. The contributions of all the factors in the maintenance productivity are not the same. Some factors contribute more to the productivity of the maintenance system than others. To obtain an overall score combining all the factors in a single productivity score, a weight for each factor is needed. Next, a methodology for assessing the weight of each factor is presented.

14.5.2 Analytic Hierarchy Process (AHP) for Determining Weights of Factors

A classical problem in decision theory is the determination of weights for a set of factors (activities) or objectives that contribute toward a common goal. The need arises in two situations. The first is when a judgment is required about the importance of a factor to a process. The second is when an aggregation of several objectives is needed. Four management science methods exist for making value and benefit assessments. These four are sorting, ranking, rating, and scoring.

Pairwise comparison is a common approach for assigning weight or ranking factors. Then, a summing procedure may be used to come up with a weight for each factor. Moody's precedence chart is an example of such a method where pairwise comparison is used and aggregated to obtain scores or ranks for factors [2]. Another approach for assigning weight is the Delphi method where a group of experts are surveyed and an interactive approach is used to obtain scores or weights for the factors under consideration.

A well-known method for weight determination, and one which is gaining popularity, is the AHP [2]. The weight should reflect the contribution of each factor in the maintenance system. In this method, a matrix of pairwise comparisons of the factors is required. The entries of the matrix indicate the strength with which one factor dominates another. The (a_{ji}) element of the matrix is defined as follows:

$a_{ij} \equiv$ Intensity of importance of factor i in comparison with factor j using the scale in Table 14.1.

$$\text{Clearly } a_{ij} = \frac{1}{a_{ji}}$$

Table 14.1 Intensity of importance scale

Intensity of importance	Definition	Explanation
1	Equal importance	Two activities contribute equally to the objective
3	Weak importance of one over another	Experience and judgement slightly favor one activity over another
5	Essential or strong importance	Experience and judgement strongly favor one activity over another
7	Demonstrated importance	An activity is strongly favored and its dominance demonstrated in practice
9	Absolute importance	The evidence favoring one activity over another is of the highest possible order of affirmation
2, 4, 6, 8	Intermediate values between the two adjacent judgements	When compromise is needed
Reciprocals of above nonzero numbers	If activity *i* has one of the above nonzero numbers assigned to it when compared with activity *j*, the *j* has the reciprocal value when compared with *i*	Set note below
Rational	Ratios arising from the scale	If consistency were to be forced by obtaining *n* numerical values to span the matrix

A scale of subjective scores from the pairwise comparison of the factors is developed using the scale in Table 14.1. In order to apply AHP to a specific weight determination problem, the following steps must be carried out.

1. Obtain the matrix of pairwise comparisons using the scale in Table 14.1.
2. Find the maximum eigen value λ_{max}. If it is close to n, the dimension of the matrix then proceeds to 3.
3. Find the eigen vector corresponding to the maximum eigen value. This vector provides the weights for each factor. The weights can be normalized to obtain weights that add up to one. The AHP method provides the means to evaluate the consistency of the weights obtained using it. It measures consistency by a consistency ratio (CR). The value of this ratio must be 10 % or less.

$$\mathrm{CR} = \frac{\lambda_{\max} - n}{n - 1} \tag{14.9}$$

The above method is used in this chapter for determining the importance of maintenance factors. An example demonstrating the use of AHP is given in Sect. 12.6. For more about AHP, see Ref. [2].

14.5.3 ABC (Pareto Chart) Analysis

This section presents ABC analysis or the Pareto chart method for identifying the most deficient factors. The ABC analysis has been utilized successively in inventory management and other areas for focusing management effort to the areas of most importance and, therefore, of greatest effect. Pareto chart or ABC analysis has been covered in Chaps. 7 and 8.

The graph obtained from the Pareto analysis will help in classifying factors or categories that account for most of the frequencies. For example, if the factors are cost centers and the frequency is the percentage of the total budget spent in each center, the ABC analysis will immediately identify the factors that are responsible for most of the cost. Actually, an ABC analysis divides the factors into three classes. The first is class A which contains between 15 and 20 % of the factors and is responsible for 70–80 % of the cost. Class B contains about 20–25 % of the factors and is responsible for 20–25 % of cost, and class C has the rest of the factors that are responsible for less than 5 % of the cost. This type of analysis helps to separate the important from the less important factors.

After obtaining the weight of each factor, a combined productivity score is calculated as follows:

$$\text{Maintenance Productivity Index} = \text{M.P.I} = \frac{\sum_{i=1}^{14} w_i S_i}{\sum_{i=1}^{14} w_i I_i} = \frac{\sum_{i=1}^{14} w_i S_i}{10} \tag{14.10}$$

where

w_i = normalized weight of factor i
S_i = score of factor I; and
I_i = maximum score of factor i and it is 10 for all factors.

An ABC analysis based on the percentage weighted deviation from the weighted ideal score is conducted to identify the most deficient factors and set priorities for factor improvement. The percentage weighted deviation for factor i (PWD_i) is calculated as follows:

$$d_i = wl_i - W_i S_i = W_i(l_i - S_i) \tag{14.11}$$

$$\text{PWD}_i = \left[\frac{d_i}{\sum_{i=1}^{14} d_i}\right] \times 100 \tag{14.12}$$

The above analysis will help to identify the factors that are responsible for the deviation from the ideal score based on their scores and importance.

14.6 Root Cause Analysis and Possible Corrective Action

In order to develop an improvement program, the deficient factors identified in the ABC analysis need to be corrected. Root cause analysis is a proper tool for identifying the causes and remedying the deficient factors.

Root cause analysis refers to the process of identifying the most basic cause of an undesirable condition or problem. It has been applied successfully in the area of total quality management. The techniques for root cause analysis include the following: change analysis, barrier analysis, events and causal factors, tree diagrams, and cause and effect diagram. The above techniques are very useful in identifying basic causes so that efforts are maximized. For a complete description of root cause analysis, see Ref. [5]. The most powerful tool of root cause analysis is the cause and effect diagram. It is a formal tool for isolating the real and root causes of a defect. The steps for constructing the diagram are shown in Chap. 8 with an illustrative example. Next, guidelines for initiating correction of deficient factors are presented.

Organization and Staffing

If organization and staffing is found deficient, possible corrective actions are the following:

(a) Establish a suitable supervisor–worker ratio.
(b) Establish a suitable planner–worker ratio.
(c) Develop a specific job description.
(d) Develop the best mix between centralized and area maintenance.
(e) Establish good support functions.
(f) Use a written labor control policy.

Labor Productivity

If labor productivity is categorized as class A or B from the Pareto chart, the following are possible corrective actions:

(a) Establish a system for collecting data on actual time spent on the jobs by craftsmen.
(b) Relate actual time spent performing a job to its standard time.
(c) In the maintenance monthly report, calculate the craft productivity index (CPI) using the following formula:

$$\text{CPI} = \frac{\text{Standard time}}{\text{Actual time}} * \text{Utilization}$$

Utilization is the percentage of time in which craftsmen are gainfully occupied, and can be obtained using work sampling.
(d) Review backlog by crafts

Management Training

If management training is categorized as class A or B, a training program for management must be launched and must be updated yearly. The program must include productivity, project management, and supervisory skill training.

Planner Training

The possible corrective actions for planner training if categorized as class A or B are as follows:

(a) The planner should be the most experienced person.
(b) The planner must coordinate with all maintenance areas.
(c) The planner must utilize modern planning and scheduling techniques.
(d) The planner must be trained in new analytical methods in planning and scheduling.

Craft Training

The possible corrective actions for craft training are as follows:

(a) Establish a yearly training program for all craftsmen.
(b) Establish a craft certification program.

Motivation

If motivation is categorized as class A or B, possible corrective actions are as follows:

(a) Make the worker feel part of the company.
(b) Establish an excellent reward policy.
(c) Reduce annual turnover and review the turnover causes on a half-yearly basis.
(d) Conduct an informal survey to see craft morale.

Management and Budget Control

If management and budget control is found to be in class A or B, the possible corrective actions are:

(a) Establish guidelines for cost and budget estimation.
(b) Analyze utilization using the percentage of downtime.
(c) Establish a monthly maintenance report.
(d) Establish group inspection programs.
(e) Use the concept of activity-based costing in cost allocation.

Work Order Planning and Scheduling

If work order planning and scheduling is found deficient and is classified as class A or B, the possible corrective actions are the following:

(a) Design a simple work order.
(b) Coordinate with other departments in the plant.
(c) Computerize the work order system.

(d) Develop job standards.
(e) Develop standard jobs.

Facilities

If the item about facilities is found to be in class A or B, possible corrective actions are the following:

(a) Reorganize the location of maintenance facilities.
(b) Enhance the practice of safety.
(c) Establish preventive maintenance for major equipment items.
(d) Establish a program for tool custody.
(e) Establish equipment availability reports.

Stores, Material, and Tool Control

If stores, material, and tool control are categorized as class A or B, possible corrective actions are the following:

(a) Establish a catalog for spares on hand.
(b) Establish a list of approved vendors.
(c) Analyze maintenance delays due to unavailability of parts.
(d) Review and update inventory policy regularly.
(e) Use an ABC analysis to establish material requirements and reorder levels.

Preventive Maintenance and Equipment History

If preventive maintenance and equipment history is found to be in class A or B, possible corrective actions are the following:

(a) Establish a record of repair history for major equipment items.
(b) Cover essential equipment by preventive maintenance.
(c) Review preventive maintenance frequency and reporting.
(d) Assign a person to be responsible for preventive maintenance planning, scheduling, and control.

Engineering and Condition Monitoring

If engineering and condition monitoring is categorized as class A or B, possible corrective actions are the following:

(a) Establish a system for diagnostic routines.
(b) Analyze equipment downtime every six months.
(c) Control downtime of equipment using predictive maintenance.

Work Measurement and Incentives

If work measurement and incentives are found to be deficient, possible corrective actions are the following:

(a) Establish job standards.
(b) Review backlog records to identify causes.
(c) Balance workload and manpower requirements.
(d) Review manpower requirements yearly.

Information System

If the information system is categorized as class A or B, possible corrective actions are the following:

(a) Identify maintenance requirements.
(b) Establish computer support for various maintenance functions.
(c) Establish a better system for data collection.
(d) Develop an information system which meets the requirements in item (a).

14.7 A Case Study for Maintenance Productivity Improvement

The plant to which this plan was applied is located in the Eastern Province of Saudi Arabia. The scoring of maintenance factors was done by the maintenance department. It is usually advisable to have two or three experts in the maintenance area in addition to two to three persons from the maintenance department when scoring and making pairwise comparison steps. The ideal and the actual factor scores are given in Table 14.2.

The pairwise comparison was done comparing the importance of factors to the productivity of the maintenance system using the scale in Table 14.1. The pairwise comparison matrix is given in Table 14.3. International Mathematical Software Library (ISML) was used to obtain the eigen vector that corresponds to the largest eigen value. Then, the vector was normalized and the weight of each factor was obtained. These are given in Table 14.4.

To identify factors contributing most to the deviation from the ideal score, an ABC analysis was conducted and is shown in Table 14.5. This shows the factors

Table 14.2 Factor scores for the plant

Factor number	Score (Si)	Ideal score (Li)	Factor number	Score (Si)	Ideal score (Li)
1	7.00	10	8	9.67	10
2	6.20	10	9	6.70	10
3	6.75	10	10	7.82	10
4	7.33	10	11	7.80	10
5	7.25	10	12	6.00	10
6	8.00	10	13	3.00	10
7	7.14	10	14	8.00	10

Table 14.3 Pairwise comparison matrix for the fourteen factors

Factor	1	2	3	4	5	6	7	8	9	10	11	12	13	14
1	1	4/9	4/6	4/2	4/9	4/8	1	4/5	4/3	4/5	1	4/7	1	4/6
2	9/4	1	9/6	9/2	1	9/8	9/4	9/5	9/3	9/5	9/4	9/7	9/7	9/6
3	6/4	6/9	1	6/2	6/9	6/8	6/4	6/5	6/3	6/5	6/4	6/7	6/4	1
4	2/4	2/9	2/6	1	2/9	2/8	2/4	2/5	2/3	2/5	2/4	2/7	2/4	2/6
5	9/4	1	9/6	9/2	1	9/8	9/4	9/5	9/3	9/5	9/4	9/7	9/4	9/6
6	8/4	8/9	8/6	8/2	8/9	1	8/4	8/5	8/3	8/5	8/4	8/7	8/4	8/6
7	1	4/9	4/6	4/2	4/9	4/8	1	4/5	4/3	4/5	1	4/7	4/1	4/6
8	5/4	5/9	5/6	5/2	5/9	5/8	5/4	1	5/3	1	5/4	5/7	5/4	5/6
9	3/4	3/9	3/6	3/2	3/9	3/8	3/4	3/5	1	3/5	3/4	3/7	3/4	3/6
10	5/4	5/9	5/6	5/2	5/9	5/8	5/4	1	5/3	1	5/4	5/7	5/4	5/6
11	1	4/9	4/6	4/2	4/9	4/8	1	4/5	4/3	4/5	1	4/7	1	4/6
12	7/4	7/9	7/6	7/2	7/9	7/8	7/4	7/5	7/3	7/5	7/4	1	7/4	7/6
13	1	7/9	4/6	4/2	4/9	4/8	1/4	4/5	4/3	4/5	1	7/4	1	4/6
14	6/4	6/9	1	6/2	6/9	6/8	6/4	6/5	6/3	6/5	6/4	6/7	6/4	1

Table 14.4 Factor weights from AHP

Factor number	Normalized weight	Factor number	Normalized weight
1	0.52	8	0.065
2	0.117	9	0.039
3	0.078	10	0.065
4	0.026	11	0.052
5	0.117	12	0.091
6	0.104	13	0.052
7	0.063	14	0.078

sorted in decreasing percentage deviation order and the cumulative deviation. Factors, 2, 13, 12, and 5 are in class A, and an improvement program must be launched for these factors to obtain the best overall improvement.

14.8 Benchmarking and World-Class Maintenance

Although benchmarking as a management approach to continuous improvement began in the 1980s, its roots go back to the time when craftsmen etched a permanent mark on their workbench to ensure their work in progress was measured against a proper consistent standard. Modern benchmarking has its roots in total quality management, where corporations seeking to gain recognition as a total quality company, through awards such as the Malcolm Baldrige National Quality Award in the USA or the PM Award from the Japan Institute of Plant Maintenance,

Table 14.5 Tabular ABC analysis

Factor number	Percentage weighted deviation	Cumulative percentage weighted deviation	Factor number	Percentage weighted deviation	Cumulative percentage weighted deviation
2	0.152	0.152	1	0.053	0.785
13	0.125	0.277	14	0.053	0.838
12	0.125	0.402	10	0.048	0.886
5	0.110	0.512	11	0.039	0.925
3	0.087	0.599	4	0.024	0.949
6	0.071	0.670	8	0.008	1.00
7	0.062	0.732			

toured other companies which had won these awards to see how they had accomplished it. Xerox Corporation pioneered competition benchmarking, and a comprehensive definition based on their original work is now generally accepted.

Benchmarking is a continuous systematic process for evaluating the products, services, and work process of organizations that are recognized as representing best practices for the purpose of organizational improvement.

Selection of benchmarking partners can come from other departments or plants within the organization (internal), direct competitors, the general industry sector, or from any enterprise regardless of sector, which performs the function or process under review better than most others. Therefore, an automotive parts stamping plant wishing to benchmark predictive maintenance may look at sister plants in that company's other locations, other competitive auto parts stamping plants, general metal fabricators in the industry or a petrochemical refinery, and nuclear power plant or airline, each of which has compelling reasons to be very good at predictive maintenance.

Benchmarking, as a continuous improvement methodology, looks at both the metrics and the processes to achieve these metrics. The methodology begins with a thorough understanding of the performance of the host plant and those processes which are elemental to the success of the plant. This will provide the scope of what should be benchmarked. In maintenance management, an assessment is made of ten areas, and the three or four most critical are candidates for the benchmarking exercise. A team comprising maintenance, engineering, materials, and operations personnel examines the following:

1. long-range planning and improvement initiatives
2. organization and manpower planning
3. work planning, scheduling, and control
4. purchasing, stores, and inventory control
5. preventive and predictive maintenance
6. performance measures and statistics
7. information management and systems
8. reliability engineering
9. maintenance/production interface and communication
10. management and administrative processes

Performance measures for maintenance management have been available for some time in several industry sectors, e.g., airlines, power generation, oil/gas/petrochemical refining, and pulp/paper processing; however, documentation of the activities in the above ten processes to determine how these performance metrics have been achieved must be done through competitive benchmarking.

Benching the process for, say, developing a cost-effective preventive and predictive maintenance (PM) program involves understanding the causal relationship between this process and others, and their performance statistics. A solid PM program reduces breakdowns, emergencies, urgent work, overtime, manpower costs, parts inventory costs and increases availability, reliability, process efficiency and equipment operating precision, and thereby product quality.

A world-class PM program uses analytical techniques such as failure mode effect and criticality analysis (FMECA)), fault tree analysis (FTA), root cause analysis; and routine methods of monitoring equipment performance such as vibration monitoring, oil analysis, thermography, equipment history trend analysis, and expert diagnostic systems.

The adaptation of the best practices in all areas to the circumstances and key success factors of an individual operation accomplishes achieving world-class maintenance.

14.9 Maintenance Process Re-engineering

Hammer [6] first articulated process re-engineering concepts in an article entitled "Don't Automate, Obliterate". The contention was that many business processes are overly complicated, inefficient, and with so many non-value-adding activities, that to simply automate them with information technology solutions will not substantively improve performance. He advocated starting over with a new, streamlined business process that addresses the most common transaction with the fewest steps and that concentrates on adding value to the customer. What adds value is high quality, speed, and service at a low cost and acceptable risk.

Historically, maintenance has been managed as a support or service function, with the sole responsibility to prevent or repair breakdowns. Production personnel operated the equipment, stores personnel ran the materials, engineering personnel designed or provided the specifications for the equipment, and each department paid attention to their functional "silo." With internal processes and performance measures geared to optimize each silo, the overall process of providing service to the customer and generating profit was often suboptimized. For example, an engineer may meet the specifications exactly with a "one off" design and therefore meet his performance target, but the unique spares and maintenance procedures, as well as the cost, will make the achievement of the objectives of the other departments more difficult.

When viewing equipment maintenance as a business process rather than a function or a department, we see that tradesmen have an important role, but so too do the operator, stores clerk, and engineer. Re-engineering the maintenance process involves the use of several techniques to help understand the current process, its costs, and its value-adding components and to help set the vision and performance targets for the new process.

14.9.1 Process Analysis

The first technique is called process analysis and comprises the following steps:

1. Set the process boundaries.
2. Map the core process flow through each step and responsibility.
3. Determine the inputs, outputs, and other performance measures.
4. Identify the key suppliers, customers, and managers.

For the emergency work request process, the process analysis would be as shown in Table 14.6.

The input is the identification of a failure or incipient failure, the output is the repair of the failure, and some key measures are the following:

- response time of the tradesman to arrive at the point of failure
- cycle time from identifying the failure to closure
- cost of the time for the participants in the process
- volume of emergency requests in a time period
- the material costs

With the process mapped, it may be more easy to identify the sources of inefficiencies, such as the following: bureaucracy, duplication, complexity, lack of

Table 14.6 Emergency work request process

Process steps	Operator	Operator foreman	Maintenance foreman	Tradesman	Maintenance clerk	Maintenance planner
Identify	⊠					
Notify	⊠					
Notify		⊠				
Notify			⊠			
Notify				⊠		
Inspect				⊠		
Do				⊠		
Report				⊠		
Document			⊠			
Enter					⊠	
Analyze						⊠
Close					⊠	

technology, lack of standardization, excessive cycle times and costs, too many workers, too easy to make errors, and over design.

14.9.2 Activity-Based Costing

The second technique is called activity-based costing. In most organizations, costs are collected by plant, department or cost center, equipment, and type of expense, such as labor, materials, services, and overhead. To improve a process, the costs for that process are required. In the maintenance management process, there are usually eight to twelve activities that are important for any organization and these are as follows: emergency work, corrective work, planning, scheduling, PM development and execution, safety and security, outsourcing, skills training, procurement, inventory management, major overhaul, data management, and data and performance management.

The approach to activity-based costing consists of the following five steps:

1. Determine the key activities.
2. Determine the activity cost (labor, materials, services, overheads, etc.) and performance (cost/unit, time quality, etc.).
3. Determine the cost driver or output of the activity.
4. Trace this activity cost to the cost objectives.
5. Evaluate the cost-effectiveness and efficiency by matching the activity cost to the process objectives. An example is shown in Table 14.7.

In the above case, there is a large portion of cost for urgent work, which is not planned, and a low cost for PM work, which will help reduce the urgent work.

Table 14.7 Maintenance activity costing

Activities	Staff salaried	Hourly staff	On-site contractors	Off-site contracts	Rentals	Transport	Materials	Total
Emergency	0	50	50	0	10	10	100	220
Urgent	30	200	200	50	20	10	500	1010
Planned	100	400	400	800	50	30	500	2280
Shutdown	0	10	0	0	0	0	0	10
PM	40	100	100	400	10	10	200	860
Indirect	0	400	50	0	0	0	0	450
Minor	0	100	100	0	10	10	100	320
Standing	0	50	50	0	10	10	60	180
Total	170	1310	950	1250	110	80	1460	5330

14.9.3 Competitive Benchmarking

The third technique, competitive benchmarking, attempts to create a vision of future performance of the process and has been described in the previous section. Once the benchmarking exercise has been performed, group "brainstorming" sessions are often held to adapt the best practices to the realities of the individual organization.

14.10 Summary

In this chapter, various measures for evaluating maintenance productivity are presented. This includes input and output measures. A host of indices that provide quantitative measures for system administration, cost-effectiveness, and system efficiency are also provided. In addition, a program for accessing maintenance productivity is presented. Furthermore, the program outlines techniques and approaches for developing a continuous improvement plan for a maintenance system. The steps of the program are given in Fig. 14.2. The program has been demonstrated and tested in a process type plant and is expected to result in a 25 % improvement in maintenance productivity. The chapter is concluded with two techniques for continuous improvement. The techniques are benchmarking and process re-engineering.

Exercises

1. Define the term mean time between failures (MTBF) and suggest two ways of computing it.
2. A steel plant was scheduled to operate for 330 days last year. It had 8 unscheduled breakdowns that resulted in 240 h of downtime. The plant had 96 h of preventive maintenance during these 330 days. The plant was shutdown for 120 h due to lack of feed. Process was slow for 60 h at 80 % rate and wet feed slowed the process also for 48 h at 50 % rate.
   The production rate at 100 % capacity is 1000 tons per 24 h. In this period, 3000 tons was rejected and 5000 tons was classified as a second-quality product and sold at 50 % value.
   (a) Compute the plant availability, *A*.
   (b) Compute the process rate, PR.
   (c) Compute the quality rate, QR.
   (d) Compute overall equipment effectiveness (OEE).
   (e) Compute the productivity capacity.
   In order to improve the productivity capacity which factor should management focus on and why?

3. List the factors that constitute the maintenance scheme.
4. List the most important three factors in the audit scheme. Justify your answer.
5. Use the analytic hierarchy process (AHP) to estimate the weight of each factor listed in Question 1.
6. Why ABC analysis is needed in the continuous improvement program?
7. List possible corrective actions for the factor engineering and condition monitoring, if found deficient from the audit program.
8. Locate an organization nearby, conduct a detailed audit program, and use the approach suggested in the plan given in this chapter to suggest improvements.

References

1. Campbell JD (1995) Uptime strategies for excellence in maintenance management. Productivity Press, Portland
2. Ishikawa K (1990) Guide to quality control. In: Quality resources. White Plains, New York
3. Kelly A (1989) Maintenance and the management. Conference communication, Farnham
4. Mann L (1983) Maintenance management, 2nd edn. Lexington Books, Lexington
5. Moody P (1983) Decision making. McGraw Hill Book Company, New York
6. Niebel BW (1994) Engineering maintenance management, 2nd edn. Marcel Dekker, Inc., New York
7. Thomas S, Vargas LG (1981) The logic of priorities. Martinus Nijhoff, Boston

Further Reading

1. Duffuaa SO, Raouf A (1996) Continuous maintenance productivity improvement using structured audit. Int J Ind Eng Appl Pract 3(3):151–160
2. Duffuaa SO, Shuaib AN, Raouf A, Budair MO (1990) Evaluation of maintenance management systems. J King Saud University, vol 4 (Eng Sci) (Special issue on maintenance), pp 29–42
3. Westerkamp TA (1987) Using computer to audit maintenance productivity. In: Hartman (ed) Maintenance management. Institute of Industrial Engineering, Industrial Engineering and Management Press
4. Wilson PF, Dell H, Anderson GF (1993) Root cause analysis tool for total quality management. ASQC Quality Press, Milwaukee
5. Wireman T (1990) World class maintenance management. Industrial Press Inc., New York

Appendix A
Statistical Review

A.1 Introduction

Maintenance decisions such as when to perform preventive maintenance on equipment or how to control the maintenance backlog require information about equipment failure and maintenance load. Equipment failure time and maintenance backlog cannot be determined with certainty. The knowledge of probability and statistics is necessary to estimate probability of failure and the maintenance load. This appendix reviews some of the basic statistical definitions and concepts that are employed as tools for modeling and analysis of maintenance systems as presented in this book. For more details see, Blank [1], Walpole and Meyers [2], Montgomery and Runger [3], Law and Kelton [4], and Jardine [5].

A.2 Probabilistic Concepts

In this section, fundamental probabilistic concepts that are necessary for understanding some of the material presented in this book, are covered and demonstrated with examples.

A.2.1 A Sample Space

A sample space S is the set of all possible outcomes of a random experiment. A random experiment is a process by which random outcomes are generated. An example of a random experiment are the outcome from tossing a coin or a dice. Other phenomenon, that could be viewed, as random experiments are the number of defective units produced per day, the number of failures per year for certain equipment, the time of failure of equipment and the time to repair equipment.

S.O. Duffuaa and A. Raouf, *Planning and Control of Maintenance Systems*,
DOI 10.1007/978-3-319-19803-3

A.2.2 An Event

An event E is a subset of a sample space. For example if a dice is tossed once, the sample space, $S = \{1, 2, 3, 4, 5, 6\}$ and $E = \{1, 3\}$ is an event.

A.2.3 Equally Likely Events

Equally likely events are events which have the same chance of occurring. For example when tossing a fair dice. The events {1}, {2}, {3}, {4}, {5}, {6} are all equally likely. The events are not equally likely if the dice is unbalanced.

A.2.4 Probability

Probability is the chance that a specific event occurs under stated conditions. It is a number between 0 and 1 inclusive. A probability statement is expressed as: There is a 0.25 probability or 25 % chances that a machine fails prior to 1000 operation hours. There are three methods to obtain a probability value. These methods are explained below.

The classical or a priori method to compute probability is as follows: If an event E can occur in m ways out of a total of n equally likely possible ways, the probability $P(E)$ is

$$P(E) = \frac{\text{number of ways for } E}{\text{total number of ways}} = \frac{m}{n}$$

The frequency or a posterior method defines probability as follows: If an equally likely event E occurs m times in a total of n trials, the observed probability of E, $OP(E)$, or relative frequency f, is

$$OP(E) = \frac{\text{Number of times } E \text{ occurred}}{\text{Total trials}} = \frac{m}{n}$$

In the limit as n becomes large, the observed probability approaches the true probability of E, $P(E)$.

$$P(E) = \lim_{n \to \infty} OP(E)$$

The subjective method uses the best educated estimate or guess of the probability of an even E. This method is necessary in the absence of data.

A.2.4.1 Example

A fair dice was tossed 100 times, the table below, shows the number of times each face occurred.

Face	1	2	3	4	5	6
Frequency	18	13	16	15	18	20

Let $E = \{3, 4\}$. Use the classical and the frequency methods to compute probability of E. The classical method gives the probability of E as:

$$P(E) = 2/6 = 1/3 = 0.33$$

The frequency method estimates $P(E)$, as:

$$OP(E) = \frac{16 + 15}{100} = \frac{31}{100} = 0.31$$

It is expected as number of tosses approaches ∞, $OP(E)$ approaches 0.33.

A.2.5 Simple Laws of Probability

Let E_1 and E_2 be two events. The compliment of E_1 denoted by E_1^1 is the set of all points in S, the sample space, and are not in E_1. The union of E_1 and E_2 (denoted by $E_1 \cup E_2$), is the set of all elements (points) in E_1 or E_2. The following are two simple laws of probability

$$P\left(E_1^1\right) = 1 - P(E_1)$$
$$P(E_1 \cup E_2) = P(E_1) + P(E_2) - P(E_1 \cap E_2)$$

A.2.6 Conditional Probability

If S is the sample space and E_1 and E_2 are two subset of S. The probability of E_1 given that E_2 occurred, [denoted by $P(E_1|E_2)$] is known as the conditional probability. In other words E_2 becomes the sample space under this condition.

Let us define the intersection of two events E_1 and E_2 [denoted by $(E_1 \cap E_2)$] as the set of elements common to both E_1 and E_2, then $P(E_1|E_2)$ is

$$P(E_1|E_2) = P(E_1 \cap E_2)/P(E_2)$$

Therefore,

$$P(E_1 \cap E_2) = P(E_1|E_2)P(E_2)$$

A.2.6.1 Example

Let us consider the problem of tossing a fair dice. Let $E_1 = \{1, 2, 3\}$ and $E_2 = \{2, 3, 4\}$ compute the probability of $E_1^1, E_1 \cup E_2, P(E_1|E_2)$. The sample space for this random experiment is

$$\begin{aligned} S &= \{1, 2, 3, 4, 5, 6, \} \\ P(E_1^1) &= 1 - P(E_1) = 1 - 3/6 = 1/2 \\ P(E_1 \cup E_2) &= \frac{4}{6} \quad \text{since}, P(1, 2, 3, 4) = \frac{4}{6} \\ P(E_1|E_2) &= \frac{P(E_1 \cap E_2)}{P(E_2)} = \frac{P(2, 3)}{P(2, 3, 4)} = \frac{2/6}{3/6} = 2/3 \end{aligned}$$

A.2.7 Statistical Independence

Two events E_1 and E_2 are statistically independent if and only if

$$P(E_1|E_2) = P(E_1)$$

or

$$P(E_1 \cap E_2) = P(E_1)P(E_2)$$

Clearly E_1 and E_2 in Sect. A.I.6.1 are not independent, since

$$P(E_1) = 1/2 \neq P(E_1|E_2) = 2/3$$

A.2.8 Population and Samples

A population is the entire selection or universe of the characteristic under study. It corresponds to the sample space in a random experiment. It could be finite or infinite. The true value of a population property is called a population parameter, which is usually denoted by a Greek letter μ, σ, λ, β etc.

A random sample or a sample is a finite portion of the population. A sample is used to estimate population parameters. A sample statistic or a statistic is a single number calculated from the sample data to estimate a population parameter. In other words statistic is a real valued function of the sample observation. A statistics is denoted by a Roman letter such as S, $\overline{X}$. For example $\overline{X}$ the sample mean is a common estimate for the population mean μ and S is an estimate for σ.

A.2.9 Discrete and Continuous Variables

A variable is the population characteristic under investigation and can be discrete or continuous. A discrete variable takes on only certain isolated values. For example, the number of accidents in a workshop.

A continuous variable can assume any value between two limits. An equipment time to failure is an example of a continuous variable.

A.3 Data Presentation and Analysis

It is very common to have lots of data which can be processed into information that can lead to better decision making. In this part of the appendix fundamental techniques for data presentation and analysis are presented.

A.3.1 Relative Frequency Distribution and Histogram

A relative frequency distribution is a table that presents the frequencies of the observed values or over specified classes from the range of the data. The graph of the frequencies versus the observed values or classes is called a histogram.

The maintenance schedule for small motors states that lubrication must be performed weekly (every 5 days) or more often. A maintenance planner checked the maintenance records and found that the days between lubrications varied from 3 to 20 days. The data for the days between lubrication is shown in Table A.1.

If we form classes of width 4 the following relative frequency Table A.2 is obtained

Table A.1 Days between lubrication

Days between lubrication	Number of motors	Percentage of motors
3	5	0.07
4	4	0.06
5	10	0.14
7	15	0.21
10	9	0.13
13	12	0.17
18	7	0.08
20	10	0.14

Table A.2 Relative frequency of lubrication

Class	Relative frequency
[0,4)	0.13
[4,8)	0.35
[8,12)	0.13
[12,16)	0.17
[16,20)	0.22

$[a,b)$ means the value a is included in the class and b not included.

The graph in Fig. A.1 shows the classes versus relative frequency which is the histogram for the data in Table A.1.

The sum of the relative frequencies is 1.0. We can use the relative frequency histogram for computing the probability of the time elapsed between lubrications is greater than 8 day which is the area to the left of 8 equal to 0.52. This means that, over 50 % of the motors has a period more than 8 days elapsed between lubrications.

The following are rules of thumb for determining the number of classes in a frequency distribution and the width of each class. The number of classes K should be:

1. $K = \sqrt{n}$, where n is a number of observations and also
2. $0 \leq K \leq 20$
3. Use different values of K to obtain a graph with no gaps.

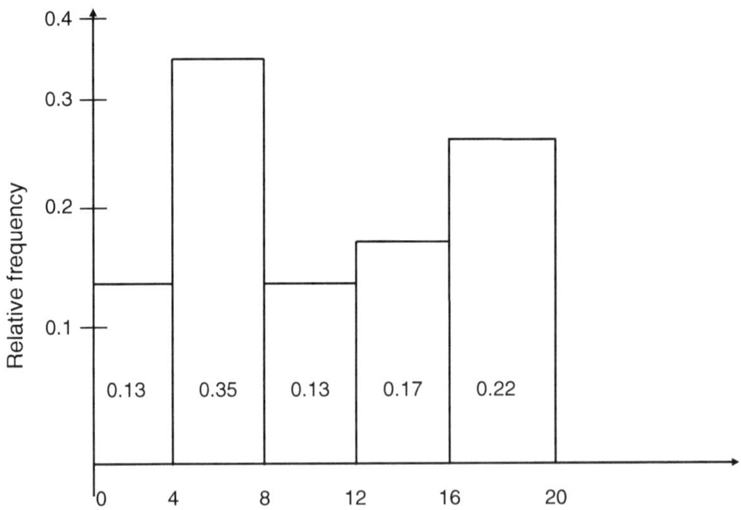

Fig. A.1 Histogram of the relative frequency of the time between lubrications

The width of the class W, is as follows:

$$W = \frac{R}{K}$$

where
$R = \max x_i - \min x_i$, i.e. the range of the data. Some refinement might be performed to the values of W or K in order to help in getting simple and natural class end points.

A.3.2 Measures of Central Tendency

The three most common measures of central tendency are the mean, median and the mode. The sample mean, $\bar{x}$ defined as

$$\bar{x} = \frac{\sum x_i}{n}$$

where Σ denote the sum of the observations, $x_1, x_2, \ldots, x_n$.

The sample medium M, defined as any value such that half of the data in the sample are greater than or equal to this value and the other half less than or equal to it.

The mode is defined as the sample value with the greatest frequency.

A.3.3 Measures of Dispersion or Variability

These measures reflect the variability in the data. The most common measures of variability are the range, R, variance S^2, standard deviation S and coefficient of variation COV.

The range R of the sample is defined as the difference between the maximum and minimum values in the sample

$$R = x_{\max} - x_{\min}$$

The variance of the data is defined as

$$S^2 = \frac{\sum_i (x_i - \bar{x})^2}{n - 1}$$

The standard deviation which is the primary measure of dispersion is:

$$S = \sqrt{\frac{\sum_i (x_i - \bar{x})^2}{n - 1}}$$

The coefficient of variation is the standard deviation per unit mean

$$COV = \frac{S}{\bar{x}}$$

A.4 Random Variables

A random variable (r.v.) is a variable that takes on its values with certain probability, and denoted by capital letter $X, Y, \ldots, Z$. Mathematically, a r.v. is a function from the sample space to the real numbers. Examples of r.vs are: the amount of rainfall in a given year, the demand for a certain product, the life of a light bulb, the amount of monthly maintenance load, the time to repair a machine and the number of accidents in a workshop in a given month. To define a r.v. precisely we need to specify its range and the distribution of the probability over the range is a discrete r.v.

A.4.1 Continuous and Discrete Random Variables

A r.v. that can take on any value on a continumum is called a continuos r.v. However, if it takes isolated values it is called discrete. The life of a light bulb is a continuous r.v. and the number of failures of a certain equipment per year.

A.4.2 Discrete Probability Mass Function

A function that defines the distribution of the probability over the range of a discrete random variable is called the probability mass function (p.m.f). The p.m.f. is given as

$$P(x) = P[X = x]$$

$P(x)$ has the following properties

$$P(x) \geq 0$$
$$\sum_x p(x) = 1$$

A.4.2.1 Example

Consider the random variable X that represents the number of requests for maintenance service per day. X can assume the integer values between 0 and 10 inclusive. The p.m.f. is defined as follows:

$$P(x) = \tfrac{x}{55} \qquad x = 0, 1, 2, \ldots, 10$$
$$= 0 \qquad \text{otherwise}$$

Notice that,

$$P(x) \geq 0 \qquad \text{for all } x \text{ in } [0,\ 10]$$

$$\sum_{x} \mathrm{P}(x) = 0 + \frac{2}{55} + \frac{3}{55} + \frac{4}{55} + \frac{5}{55} + \frac{6}{55} + \frac{7}{55} + \frac{8}{55} + \frac{9}{55} + \frac{10}{55} = 1$$

Therefore $P(x)$ satisfies the condition of a p.m.f.

A.4.3 Probability Density Function

The probability density function (p.d.f.), $f(x)$ is a function that expresses the distribution of the probability over the range of a continuous r.v. It has the following properties

$$f(x) \geq 0$$

$$\Pr[a \leq x \leq b] = \int_{a}^{b} f(x)\mathrm{d}x$$

$$\int_{-\infty}^{\infty} f(x)\mathrm{d}x = 1$$

The p.d.f. is a relative frequency histogram with infinite number of classes, each with very small width.

A.4.3.1 Example

The time to repair an equipment is equally likely between [10,20] minutes. Find the probability a repair will take less than 15 min

$$f(x) = \tfrac{1}{10}, \quad 10 \leq x \leq 20$$
$$= 0, \quad \text{otherwise}$$

Clearly,

$$f(x) \geq 0$$

$$\int_{-\infty}^{\infty} f(x)\mathrm{d}x = \int_{10}^{20} \frac{1}{10}\mathrm{d}x = \frac{1}{10}x\Big]_{10}^{20} = \frac{20}{10} - \frac{10}{10} = 1$$

$$P[X \leq 15] = \int_{10}^{15} \frac{1}{10}\mathrm{d}x = \frac{1}{10}x\Big]_{10}^{15} = \frac{15}{10} - \frac{10}{10} = 1/2$$

A.4.4 Distribution Function

The cumulative probability distribution function known as the distribution function is defined as

$$F(x) = \Pr[X \leq x]$$

For the discrete r.v. $F(x)$ will be

$$F(x) = \sum_{t \leq x} P(t)$$

and, for the continuous r.v. $F(x)$ will be

$$F(x) = \int_{-\infty}^{x} f(t)\mathrm{d}t$$

$F(x)$ can be used to determine the probability of the r.v. Between two values i.e. $P[a \leq x \leq b]$, for the continuous random variable as follows:

$$\Pr[a \leq x \leq b] = F(b) - F(a)$$

This formula can be used for the discrete case with some care about whether the interval end points are included or not.

$F(x)$ plays an important role in computing equipment reliability. For example if the time to failure T has a p.d.f. $f(t)$. Then the reliability of an equipment at time t_0, $R(t_0)$ is the probability of the equipment surviving beyond time t_0

$$R(t_0) = P[T \geq t_0] = 1 - \Pr[T \leq t_0] = 1 - F(t_0)$$

F(*t*) is very instrumental in the development of preventive maintenance policies.

A.4.5 Expected Value of a Random Variable

The expected value *E*(*x*) or the mean of a r.v. is defined as follows. For a discrete r.v.

$$\mu = E(X) = \sum_x x P(x)$$

For a continuous r.v.

$$\mu = E(X) = \int_{-\infty}^{\infty} x f(x) \mathrm{d}x$$

The expected value of any function *g*(*X*) of a r.v. denoted by *E*[*g*(*X*)] given as: For the discrete case, it is

$$E[g(X)] = \sum_x g(x) p(x)$$

For the continuous case, it is

$$E[g(X)] = \int_{-\infty}^{\infty} g(x) f(x) \mathrm{d}x$$

Also it is true that the expected value of the sum of two function of a r.v. is the sum of the expected value. In other words if *g*(*X*) and *h*(*X*) are functions of a r.v. *X*, the

$$E[g(X) + h(X)] = E[g(X)] + E[h(X)]$$

A.4.5.1 Example

Find the expected values for the r.vs in Sects. A.3.2.1 and A.3.3.1. The expected value of the r.v. in Sect. A.4.2.1 is

$$E(x) = \sum_{x=0}^{10} x P(x) = 0 + \frac{1}{55} + \frac{4}{55} + \frac{9}{55} + \frac{16}{55} + \frac{25}{55} + \frac{36}{55} + \frac{49}{55} + \frac{64}{55} + \frac{81}{55} + \frac{100}{55} = 7$$

The expected value for the r.v. in Sect. A.3.3.1 is

$$E(x) = \int_{10}^{20} x\frac{1}{10}\mathrm{d}x = \left.\frac{x^2}{20}\right]_{10}^{20} = \frac{400}{20} - \frac{100}{20} = 20 - 5 = 15$$

A.4.6 The Variance and Standard Deviation of a Random Variable

The variance of a random variable X is defined as

$$\sigma^2 = \mathrm{var}(X) = E\left[(X-\mu)^2\right] = E\left[X^2\right] - \mu^2$$

where μ is the mean of X.

For a discrete random variable

$$\sigma^2 = \sum_x (x-\mu)^2 P(x)$$

For a continuous r.v. the variance is given as

$$\sigma^2 = \int_{-\infty}^{\infty} (x-\mu)^2 f(x)\mathrm{d}x$$

The standard deviation of X, is the square root of the variance. In most real cases, μ and σ^2 are unknown, however they are usually, estimated by $\hat{X}$ and S^2.

A.5 Selected Probability Distributions

Probability distributions are essential for statistical inference and modeling random variables arising in maintenance problems such as equipment failure and time to complete a repair task. The exponential, lognormal, and Weibull are used extensively in modeling failure data. The exponential, uniform are useful for modeling the time to complete a maintenance task. The Poisson is useful for representing the number of failures in given time period. The normal, t, chi-square and F-distribution are necessary for estimation and testing various statistical hypothesis. In this part some known useful distribution are presented.

A.5.1 The Poisson Distribution

This distribution is useful for modeling the number of units arriving per unit time or number of particles per unit area. It can be used to represent the number of maintenance work orders received per day or per month or the number of accidents per unit time. The probability mass function for the Poisson r.v. is

$$P(X = x) = \frac{\lambda^x e^{-\lambda}}{x!}, \quad x = 0, 1, 2, \ldots$$

where $x!$ is the factorial function i.e. $x! = x(x-1)(x-2)\ldots1$

The Poisson distribution has one parameter λ, which equal to its mean, μ and variance, σ^2, i.e.:

$$\mu = \lambda, \quad \sigma^2 = \lambda$$

A.5.2 The Exponential Distribution

The exponential distribution represents time to failure data when the failure, is truly random, or when the failure rate is constant. Such failures usually result from a sudden excessive loading. Also it can model time to complete a service. This distribution is found to be typical for modeling the time to failure of many electronic components and pieces of industrial plants. The failure rate of such equipment is constant. The density function of the exponential distribution is

$$f(t) = \frac{1}{\lambda} e^{-x/\lambda} \quad \text{for} \quad t \geq 0$$

The mean and the variance of the distribution are: $\mu = \lambda, \quad \sigma^2 = \lambda$

A.5.3 The Uniform Distribution

The uniform distribution represents situation when every point can occur equally likely. In practice, it can be used for modeling tasks completion times.

The density function of the uniform random variable X, denoted $U[a,b]$ is

$$f(x) = \frac{1}{b-a} \quad a \leq x \leq b$$
$$= 0 \quad \text{otherwise}$$

The mean and the variance of the $U(a, b)$ r.v. is

$$\mu = \frac{a+b}{2}, \quad \sigma^2 = \frac{1}{12}(b-a)^2$$

A.5.4 Normal Distribution

The normal (Gaussian) distribution plays a central role in estimation and test of hypothesis. It can model random effects which are the consequences of a large number of small and independent r.vs. The graph of the normal distribution is bell shaped. Its density function is

$$f(x) = \frac{1}{\sigma\sqrt{2\pi}} \exp\left(\frac{-(x-\mu)^2}{2\sigma^2}\right), \quad -\infty < x < \infty$$

μ is the mean and σ is the standard deviation of the normal distribution. The normal distribution can be standardized to what is known as the standard normal with mean 0 and variance 1. The standard normal is usually denoted by Z and obtained from the general normal by the following transformation:

$$Z = \frac{X-\mu}{\sigma}$$

$$f(z) = \frac{1}{\sqrt{2\pi}} e^{\frac{-z^2}{2}}, \quad -\infty < z < \infty$$

A.5.5 The Weibull Distribution

The Weibull distribution arises in modeling failure data for equipment subject to wear. It is also used in modeling task completion time. The p.d.f. of the Weibull distribution is defined as

$$\begin{aligned} f(x) &= \alpha\beta^{-\alpha}x^{\alpha-1}e^{-\left(\frac{x}{\beta}\right)^{\alpha}} \quad && x > 0 \\ &= 0 && \text{otherwise} \end{aligned}$$

$\alpha > 0$ is a shape parameter and $\beta > 0$, is a scale parameter. The distribution function $F(x)$ is given as

$$F(x) = 1 - e^{-\left(\frac{x}{\beta}\right)^{\alpha}}, \quad x > 0$$

The mean and the variance of the Weibull distribution are:

$$\mu = \frac{\beta}{\alpha}\Gamma\left(\frac{1}{\alpha}\right)$$
$$\sigma^2 = \frac{\beta^2}{\alpha}\left[2\Gamma\left(\frac{2}{\alpha}\right) - \frac{1}{\alpha}\left[\Gamma\left(\frac{1}{\alpha}\right)\right]^2\right]$$

A.5.6 The Lognormal Distribution

The lognormal distribution represents time to complete some task. If X is lognormally distributed, than *In* X is normally distributed. The density function for the lognormal random variable,

$$\begin{aligned} f(x) &= \frac{1}{x\sqrt{2\pi\sigma^2}}\exp\left(\frac{-(\ln x-\mu)^2}{2\sigma^2}\right), & x > 0 \\ &= 0 & \text{otherwise} \end{aligned}$$

μ is a scale parameter and $\sigma > 0$ is a shape parameter. The mean and the variance of the lognormal distribution are:

$$\mu = e^{\mu+\frac{\sigma^2}{2}}$$
$$\sigma^2 = e^{2\mu+\sigma^2}(e^{\sigma^2} - 1)$$

A.5.7 Chi-square Distribution

This distribution plays a central role in statistical inferences about variability. The continuous random variable X has a chi-square distribution, with v degrees of freedom if its density function is of the form

$$\begin{aligned} f(x) &= \frac{1}{2^{v/2}\Gamma\left(\frac{v}{2}\right)}x^{v/2-1}e^{-x/2} & x > 0 \\ &= 0 & \text{otherwise} \end{aligned}$$

where v is a positive integer. The mean and the variance of the chi-square distribution are given as:

$$\mu = v$$
$$\sigma^2 = 2v$$

It has been proven that the quantity $\frac{(n-1)S^2}{\sigma^2}$ has a chi-squared distribution with $(n - 1)$ degrees of freedom and is used for developing confidence intervals and testing statistical hypothesis about σ.

A.5.8 The t-Distribution

The applications of the t-distribution are in the area of inferences on a population mean or the difference between two means. Let Z be a standard normal r.v. and V is a chi-squared r.v. with v degrees of freedom. If Z and V are independent, then the distribution of the r.v., T given by

$$T = \frac{Z}{\sqrt{V/v}}$$

is a t-distribution with v degrees of freedom. The density function of T is

$$f(t) = \frac{\Gamma[v + 1/2]}{\Gamma\left[\frac{v}{2}\right]\sqrt{\pi v}}\left(1 + \frac{t^2}{v}\right)^{-(v+1)/2} \qquad -\infty < t < \infty$$

A.5.9 F-Distribution

The F-distribution is used for testing hypothesis about several means and two variances. It is a ratio of two independent chi-squared distribution. Let U and V be two independent random variables having chi-square distribution with v_1 and v_2 degrees of freedom respectively. Then the quantity

$$F = \frac{U/v_1}{V/v_2}$$

has an F-distribution, with v_1 and v_2 degrees of freedom. The density function of the F-distribution is given as

$$f(x) = \frac{\Gamma[v_1 + v_2/2](v_1/v_2)^{v_1/2}}{\Gamma[v_1/2]\Gamma[v2/2]} \qquad \frac{x^{v_1/2} - 1}{\left(1 + \frac{v_1 x}{v_2}\right)^{(v_1+v_2)/2}}$$

$$= 0 \qquad \text{otherwise}$$

The quantity $\frac{S_1^2/\sigma_2^2}{S_2^2/\sigma_2^2}$ has an F-distribution with $n_1 - 1$, $n_2 - 2$ degrees of freedom, where n_1 is the sample size from the first population and n_2 is the sample size from the second population.

A.6 Estimation

Estimation is the process by which population parameters are estimated. A point estimate is a single-valued estimate of a population parameter. An interval estimate is a probability statement that a population parameter lies between computed values. An interval estimate has a level of confidence associated with it. The confidence level is the degree of assurance that a particular statistical statement is true under specified condition. The significance level is the degree of uncertainty about the statistical statement under the same conditions used to determine the confidence level. $1 - \alpha$ is the confidence level and α is the significance level.

A.6.1 The Central Limit Theorem

If a random sample is drawn from a population with known and finite mean μ and variance σ^2, then the quantity $Z = \frac{\overline{X}-\mu}{\sigma/\sqrt{n}}$ has a standard normal distribution. Regardless of the type of distribution the sample is drawn from, the mean of sample form a normal distribution, if the sample size is sufficiently large.

A.6.2 Interval Estimation

The $1 - \alpha$ confidence interval for the mean μ is given by the following formula

1. $\overline{x} - Z_{\alpha 1/2}\frac{\sigma}{\sqrt{n}} \le \mu \le \overline{x} + Z_{\alpha/2}\frac{\sigma}{\sqrt{n}}$, if σ is known and $Z_{\alpha/2}$ is given by $\int_{-Z_{\alpha/2}}^{Z_{\alpha/2}} \frac{1}{\sqrt{2\pi}} e^{-x^2/2} dx = 1 - \alpha$
2. $\overline{x} - t_{\alpha 1/2}\frac{S}{\sqrt{n}} \le \mu \le \overline{x} + t_{\alpha/n}\frac{S}{\sqrt{n}}$ if σ unknown and $t_{\alpha/2}$ is defined as:

$$\int_{-t_{\alpha/2}}^{t_{\alpha/2}} f(t)dt = 1 - \alpha$$

where $f(t)$ is the density function for the t-distribution. Tables for the standard normal and the t-distributions are available in many statistics books giving the values of $Z_{\alpha/2}$ and $t_{\alpha/2}$, examples of such books are [2,3].

A.6.3 Sample Size

The sample size to assure that $\overline{x}$ (estimate for μ) is within maximum absolute error e from μ with $1 - \alpha$ confidence, assuming σ is known is

$$n = \left(\frac{Z_{\alpha/2}\sigma}{e}\right)^2$$

for the use of this formula see Chap. 4.

A.6.4 The Maximum Likelihood Estimator

The maximum likelihood estimate (M.L.E) of a population parameter is the value of the parameter in terms of the observation that maximizes the log of the likelihood function. The likelihood function is defined as:

$$L(\theta_1, \theta_2, X_1, \ldots, X_n) = \prod_{i=1}^{n} f(\theta_1, \theta_2, X_i)$$

where θ_1, θ_2 are the population parameters, X_i, $i = 1,\ldots, n$ are n observations from the population and $f(\theta_1, \theta_2, x)$ is the density or mass function of the r.v.

A.6.4.1 Example

Derive the maximum likelihood estimate for the exponential distribution parameter

The density function is given as

$$f(\lambda, x) = \frac{1}{\lambda} e^{-x/\lambda}, \quad x > 0$$

The likelihood function for a random sample (X_1, X_2,..., X_n) is:

$$L(\lambda, X_1, X_2, \ldots, X_n) = \prod_{i=1}^{n} \frac{1}{\lambda} e^{\frac{-X_i}{\lambda}} = \frac{1}{\lambda^n} \exp\left(\frac{-\sum_{i=1}^{n} X_i}{\lambda}\right)$$

$$\ln L(\lambda, X_1, X_2, \ldots, X_n) = -n \ln \lambda - \frac{\sum_{i=1}^{n} X_i}{n}$$

The M.L.E. estimator for λ maximizes ln $L(\lambda, X_1, X_2,\ldots, X_n)$

$$\frac{dL(\lambda, x_1, x_2, \ldots, x_n)}{d\lambda} = \frac{-n}{\lambda} + \frac{\sum_{i=1}^{n} X_i}{\lambda^2} = 0$$

$$\hat{\lambda} = \frac{\sum_{i=1}^{n} X_i}{n} = \overline{X}$$

To check if $\hat{\lambda}$ is maximum we find the second derivative

$$\frac{d^2L}{d\lambda_2}(\lambda, X_1, X_2, \ldots, X_n) = \frac{n}{\lambda^2} - 2\frac{\sum X_i}{\lambda^3}$$

Substitute $\hat{\lambda} = \overline{X}$ in the second derivative we get

$$\frac{\partial^2 L}{\partial \lambda_2}(\hat{\lambda}, X_1, X_2, \ldots, X_n) = \frac{n^3}{\left(\sum X_i\right)^2} - 2\frac{n^3}{\left(\sum X_i\right)^2}$$
$$= n^3\left[-\frac{1}{\left(\sum_{i=1}^n X_i\right)}\right] < 0$$

Since, $\left(\sum_{t=1}^n X_i\right) > 0$. Therefore $\hat{\lambda} = \overline{X}$ maximizes ln $L(\lambda_1, X_1, X_2, \ldots, X_n)$ and hence the M.L.E for λ is $\overline{X}$.

A.6.5 The Moment Method for Estimation

The moment method estimates the population parameters by equating the moments of the population to the moments of the data. If we have two parameters we equate the first moment of the data to the first moment of the population and the second moment to the second moment. Then solve the resulting equations.

The first moment of a population is $E(X) = \mu$ and nth population moment is given by $E(X - \mu)^2$. The first moment of the data is $\overline{X}$ and the nth moment of the data is given by $\frac{\sum_{i=1}^n (X_i - \overline{X})^n}{n}$.

A.6.5.1 Example

Estimate the parameters of the lognormal distribution using the method of the moments. The method of the moment equates the sample moments with the population moments. Let $X_1, X_2, \ldots, X_n$ be a random sample from the population. Then we have

$$E(X) = e^{\mu+\sigma^2/2} = \frac{\sum_{i=1}^n X_i}{n} = \overline{X}$$

$$\text{Variance}\,(X) = e^{2\mu+\sigma^2}\left(e^{\sigma^2} - 1\right) = \frac{\sum \left(X_i - \overline{X}\right)^2}{n} = S^2$$

Take ln for both equations we obtain

$$\mu + \frac{\sigma^2}{2} = \ln \overline{X}$$
$$2\mu + \sigma^2 + \ln(e^{\sigma^2} - 1) = \ln S^2$$

Multiply the first equation by 2 and subtract it from the second equation, we get

$$\ln(e^{\sigma^2} - 1) = \ln S^2 - \ln \overline{X}^2 = \ln \frac{S^2}{\overline{X}^2} = \ln\left(\frac{S}{\overline{\overline{X}}}\right)^2$$

Then,

$$e^{\sigma^2} - 1 = \left(\frac{S}{\overline{\overline{X}}}\right)^2$$
$$e^{\sigma^2} = \left(\frac{S}{\overline{\overline{X}}}\right)^2 + 1$$

Therefore,

$$\hat{\sigma}^2 = \ln\left(\left(\frac{S}{\overline{\overline{X}}}\right)^2 + 1\right) = \ln\left(\frac{S^2 + \overline{X}^2}{\overline{X}^2}\right)$$

Use the first equation to estimate μ

$$2\mu + \sigma^2 = \ln \overline{X}^2$$
$$2\mu = \ln \overline{X}^2 - \hat{\sigma}^2 = \ln \overline{X}^2 - \ln\left(\frac{S^2}{\overline{X}^2} + 1\right)$$
$$= ln \frac{\overline{X}^2}{S^2/\overline{X}^2 + 1}$$

Therefore

$$\hat{\mu} = 1/2 \ln\left(\frac{\overline{X}^4}{S^2 + \overline{X}^2}\right)$$

The parameters μ and σ^2 are estimated using the moment method as:

$$\hat{\mu} = 1/2 \ln \frac{\overline{X}^4}{S^2 + \overline{X}^2}$$

$$\hat{\sigma}^2 = \ln\left(\frac{S^2 + \overline{X}^2}{\overline{X}^2}\right)$$

A.7 Test of Hypothesis

A hypotheses is a statement or an assumption about one, or more populations. Hypothesis can be formulated about means, variances, differences of means or pdf forms. For test of hypothesis we use samples and knowledge of statistical distributions to make inferences about the populations from which the samples are drawn. In each test there will be two hypothesis. One is known as the null hypothesis (H_0), and the other is called the alternative (H_1). An example of a hypothesis formulation is

H_0 $X_1, X_2, \ldots, X_n$ comes from an exponential distribution
H_1 $X_1, X_2, \ldots, X_2$ do not come from an experimental distribution

where X_1, X_2,..., X_n is a random sample.

In testing statistical hypothesis two types of errors might occur. Rejecting a true H_0 is called type I error denoted by α, and accepting a false H_0 is known as type II error denoted by β. Both types of error cannot be minimized simultaneously. Minimizing α increases β and vice versa. Statisticians fix a level for α called the level of significance and use the test statistics that maximizes $1 - \beta$ the power of the test.

A.7.1 Steps for Conducting a Statistical Test of Hypothesis

The steps for performing a test of hypothesis are:

1. Formulate and state the hypothesis: the null hypothesis H_0 and the alternative H_1.
2. Decide on an appropriate level of significance α (i.e. the probability of a reject a true H_0).
3. Use the right test statistic for the hypothesis stated e.g. the standard normal, t-test Chi-square test or an F-test.
4. Determine the rejection criteria (critical region) using an appropriate distribution. Usually the values that determine the region are given in statistical tables based on steps 2 and 3.

5. Take a sample from the population and compute the value of the test statistics. If the value falls in the rejection region reject H_0. Otherwise do not reject (accept).

There are several test statistics for single mean, two means, single variance, two variance and goodness of fit.

The tests for a single mean are:

1. $Z = \sqrt{n}\frac{(\overline{X}-\mu)}{\sigma}$ when σ is known. The quantity Z is a standard normal by the central limit theorem.
2. $t = \sqrt{n}\frac{(\overline{X}-\mu)}{S}$ when σ is unknown and estimated as S. The quantity $\sqrt{n}\frac{(\overline{X}-\mu)}{S}$ has a t-distribution with $n - 1$ degrees of rejection.

The test for the difference of two means are:

1. $Z = \frac{(\overline{X}_1-\overline{X}_2)-(\mu_1-\mu_2)}{\sqrt{\sigma_1^2/n_1+\sigma_2^2/n_2}}$, when σ_1 and σ_2 are unknown. The quantity Z has a standard normal distribution.
2. $t = \frac{(\overline{X}_1-X_2)-(\mu_1-\mu_2)}{S_p\sqrt{1/n_1+1/n_2}}$, when σ_1 and σ_2 are unknown, but equal (unknown variances but equal), where,
   $S_p^2 = \frac{(n_1-1)S_1^2+(n_2-1)S_2^2}{n_1+n_2-2}$ The quantity t has a t-distribution with $n_1 + n_2 - 2$ degrees of freedom.
3. $t = \frac{(\overline{X}_1-\overline{X}_2)-(\mu_1-\mu_2)}{\sqrt{S_1^2/n_1+S_2^2/n_2}}$ The quantity t has a t-distribution with v. degrees of freedom given as: $v = \frac{\left(S_1^2/n_1+S_2^2/n_2\right)^2}{\frac{\left(S_1^2/n_1\right)^2}{n_1-1}+\frac{\left(S_2^2/n2\right)^2}{n_2-1}}$ and used in the case variances unknown and unequal.

The test statistics for the variance σ^2 equal to a specific value $\left(\text{i.e. } \sigma_1^2 = \sigma_0^2\right)$ is $X^2 = \frac{(n-1)S^2}{\sigma_0^2}$ has a X^2-distribution with $(n - 1)$ degrees of freedom.

The test statistics for the equality of two population variances (i.e. $\sigma_1^2 = \sigma_2^2$) is: $F = \frac{S_1^2/\sigma_1^2}{S_2^2/\sigma_2^2}$ has an F-distribution with $n_1 - 1$ and $n_2 - 1$ degrees of freedom.

The test statistics for the goodness of fit for a set of data to a distribution with p. d.f. $f(t)$ is:

$X^2 = \sum_{i=1}^{k} \frac{(O_i-E_i)^2}{E_i}$ has a X^2-distribution with $k - 1$ degrees of freedom, where k is the number of classes for the histogram of the data set. O_i is the observed frequency and E_i is the expected or theoretical frequency

$E_i = \int_{C_{i-1}}^{C_i} f(t)\mathrm{d}t, c_{i-1}$ is the lower limit of class i and C_i is the upper limit of class i.

A.7.1.1 Example

It has been hypothesized that the time to complete a maintenance inspection task is uniformly distributed between 15 and 30 min. To test this hypothesis a sample of size 100 has been taken from the historical records and its frequency distribution is shown below. The class is of the from $[a,b)$, a included b is excluded.

Class	15–18	18–21	21–24	24–27	27–30
Frequency	22	18	16	23	21

We follow the steps stated in Sect. A.6.1

1. H_0: The data come from $U[15,30]$
   H_1: The data do not come from $U[16,30]$
2. $\alpha = 0.05$
3. The test statistics is $X^2(k-1) = \sum_{i=1}^{k} \frac{(O_i - E_i)^2}{E_i}$
4. Determine the critical region. The critical value is obtained from the X^2 table with 4-degrees of freedom. $X^2_{0.05} = 9.49$. If the calculated value is greater than 9.49, H_0 is rejected.
5. Using the sample and the frequency table we compute the X^2 value using the formula in (3) above.

Class	15–18	18–21	21–24	24–27	27–30
O_i	22	18	16	23	21
E_i	20	20	20	20	20

The calculated chi-square (X^2) value is

$$X^2 = \frac{(22-20)^2}{20} + \frac{(18-20)^2}{20} + \frac{(16-20)^2}{20} + \frac{(23-20)^2}{20} + \frac{(21-20)^2}{20}$$
$$= \frac{4}{20} + \frac{4}{20} + \frac{16}{20} + \frac{9}{20} + \frac{1}{20} = \frac{34}{20} = 1.7$$

Since the calculated value is less than the table value. The decision is not to reject H_0 i.e. the data does not present any evidence to reject H_0.

A.8 Distribution Fitting

In many practical problems information about equipment failure distribution or maintenance tasks completion times distribution may be needed for maintenance decision making. Data on failure or tasks completion times may be available from historical records. In many situation it might be necessary to identify distributions that fit the data. This section provide the steps for fitting a distribution to a set of data. The distributions are needed for reliability analysis, finding optimal preventive maintenance policies, or simulation studies to determine the optimal manning level.

A.8.1 Steps for Distribution Fitting

1. Develop a histogram for the data or line graph.
2. From the shape of the histogram or the line graph, hypothesize a family of distribution from which the data might have come. This depends on the shape of the histogram and the shape of the distribution density curve. It could be normal, Weibull, lognormal, exponential, uniform or Poisson, etc.
3. Estimate the parameters of the hypothesized family using the maximum likelihood method or the moment method.
4. Test the goodness of fit that the data follows the distribution with the estimated parameters. Use the χ^2-test in Sect. A.6.2.

A.9 Least Square Regression

Regression is a mathematical technique that is employed to determine the line or curve that best fits a set of observed data by minimizing the deviations between the observed data and the regression equation values.

If the deviation is measured in terms of the squared error (residual) and the sum of squared deviations is minimized, then the criterion used for minimizing the errors is known as the least-squares principle. This principal is mostly used in determining the regression equation. In symbols, if we define,

$$\begin{aligned}\hat{x}_i &= \text{Estimated values by the regression equation}, \quad i = 1, 2, \ldots, n\\ x_i &= \text{Observed values}, \quad i = 1, 2, \ldots, n\\ \in_i &= \text{Error or deviation between } \hat{x}_i, \text{ and } x_i\end{aligned}$$

then,

$$\varepsilon_i = x_i - \hat{x}_i$$

The least square method minimizes

$$\text{SS} = \sum_{i=1}^{n} \varepsilon_i^2 = \sum_{i=1}^{n} (x_i - \hat{x}_i)^2$$

If a straight line relation is assumed to be the best between t and x, then the mathematical models for it is

$$x = a + bt$$

The best estimates for a and b are obtained using the least squares in the following manner

Minimize

$$\mathrm{SS} = -\sum_{i=1}^{n} (x_i - a - bt_i)^2$$

The minimization is performed using, partial differentiation of the above equation to obtain the following equations, known as the normal equations:

$$\frac{\partial \mathrm{SS}}{\partial a} = -2\sum_{i=1}^{n} (x_i - a - bt_i) = 0$$

$$\frac{\partial \mathrm{SS}}{\partial b} = -2\sum_{i=1}^{n} t_i(x_i - a - bt_i) = 0$$

Solving the above equations simultaneously, the best estimates for a and b are obtained as

$$\hat{b} = \frac{n\sum_i^n t_i x_i - (\sum t_i)(\sum x_i)}{n\sum_{i=1}^n t_i^2 - (\sum_{i=1}^n t_i)^2}$$

$$\hat{a} = \bar{x} - b\bar{t}$$

$$\text{where } \bar{x} = \frac{\sum_{i=1}^n x_i}{n}, \quad \bar{t} = \frac{\sum_{i=1}^n t_i}{n}.$$

Appendix B
Reliability and Failure Analysis

B.1 Reliability Function

The reliability function, sometimes called the survival function is a function complementary to the distribution function defined in Kapur and Lamberson [6] as

$$R(t) = \int_t^\infty f(t)\mathrm{d}t = 1 - F(t)$$

The reliability function for some known distribution is given below:

(a) Exponential with parameter λ, $R(t) = \exp\left(\frac{-t}{\lambda}\right)$
(b) Uniform between a and b, $U[a, b]$, $R(t) = 1 - \frac{1}{b-a}(t - a)$
(c) Weibull with shape parameter α and scale parameter β

$$R(t) = \mathrm{e}^{-(x/\beta)^\alpha}$$

B.2 The Expected Life

If $f(t)$ is the p.d.f. of the time to failure, the expected life for a component is $E(t) = \int_0^\infty tf(t)\mathrm{d}t = \int_0^t R(t)\mathrm{d}t$

B.3 Failure Rate Function

The failure rate function is the probability that the equipment will fail in the next interval $(t, t + \delta t)$, provided it survived up to t. It is a conditional probability

S.O. Duffuaa and A. Raouf, *Planning and Control of Maintenance Systems*,
DOI 10.1007/978-3-319-19803-3

$$r(t)\delta t = \frac{\int_t^{t+\delta t} f(t)\mathrm{d}t}{\int_t^{\infty} f(t)\mathrm{d}t} = \frac{F(t+\delta t) - F(t)}{1 - F(t)}$$

divide by δt and let $\delta t \to 0$

$$\begin{aligned} r(t) &= \lim \frac{F(t+\delta t) - F(t)}{\delta t} / 1 - F(t) \\ &= \frac{f(t)}{1 - F(t)} = \frac{f(t)}{R(t)} \end{aligned}$$

$r(t)$ is known as the instantaneous failure rate or the hazard function. Following are the failure rate function for some known distributions

(a) Exponential with parameter λ: $r(t) = 1/\lambda$, constant failure rate
(b) Uniform, $U(a, b)$: $r(t) = \frac{1}{b-t}$, increasing failure rate, provided, a $a \le t \le b$
(c) Weibull with shape parameter α and scale parameter β: $r(t) = \frac{\alpha}{\beta}\left(\frac{t}{\beta}\right)^{\alpha - 1}$ increasing failure rate.

B.4 Expected Number of Failures

The number of failures in an interval $(0, t_p)$, denoted by $N(t)$ is a r.v. The expected number of failures in the interval $(0, t_p)$ denoted by $H(t_p)$

$$H(t_p) = E[N(t)] = \int_0^{t_p} r(t)\mathrm{d}t.$$

Appendix C
Optimization Methods

Optimization methods are necessary for optimal maintenance decision making. They are needed for developing optimal preventive maintenance policies, and optimal tasks scheduling. In Chap. 3 the golden section method is presented. In this appendix the basics of optimization is presented together with two algorithms. For more on optimization method see Bazaraa et al. [7], and Luenberger [8].

C.1 Absolute value and norm

The quantity $|x - y|$ denotes the absolute value between two real values x, y. If X and Y are vectors, then $||X - Y||$ denote the norm, the distance between X and Y defined as

$$\sqrt{\sum_{i=1}^{n} (x_i - y_i)^2}$$

C.2 Derivate and partial derivative

The derivative of a function of a single variable $f(t)$ is defined as:

$$\frac{\mathrm{d}f}{\mathrm{d}t} = f'(t) = \lim_{\delta t \to 0} \frac{f(t + \delta t) - f(t)}{\delta t}$$

The partial derivative for a function of several variables is defined as follows:

$$\frac{\partial f}{\partial x_i} = \lim_{\delta x_i \to 0} \frac{f(x_1, x_2, \ldots, x_i + \delta x_i, x_{i+1}, \ldots, x_n) - f(x_1, x_2, \ldots, x_n)}{\delta x_i}$$

It can be viewed as the derivative with respective to x_i, keeping all other variables fixed.

S.O. Duffuaa and A. Raouf, *Planning and Control of Maintenance Systems*,
DOI 10.1007/978-3-319-19803-3

C.3 The Gradient

The gradient is the vector of partial derivatives

$$\nabla f(x) = \begin{pmatrix} \frac{\partial f}{\partial x_1} \\ \frac{\partial f}{\partial x_2} \\ \vdots \\ \frac{\partial f}{\partial x_n} \end{pmatrix}$$

C.4 Hessian Matrix

The Hessian matrix is the matrix of cross-partial derivatives

$$\begin{vmatrix} \frac{\partial^2 f}{\partial x_1^2} & \frac{\partial^2 f}{\partial x_1 \partial x_2} & & \frac{\partial^2 f}{\partial x_1 \partial x_n} \\ \vdots & & \frac{\partial^2 f}{\partial x_i \partial x_j} & \quad \cdots \\ \vdots & & & \\ \frac{\partial^2 f}{\partial x_1 \partial x_n} & \cdots & \cdots & \frac{\partial^2 f}{\partial x_n^2} \end{vmatrix}$$

C.5 A Stationary Point

X^* is a stationary point if and only if

$$\nabla f(X^*) = 0$$

In case of a function of a single variable $f'(x^*) = 0$.

C.6 Local Minimum

A point X^* is a local minimum if it satisfies the following necessary and sufficient conditions

Necessary conditions: $\nabla f(X^*) = 0$
Sufficient conditions: $H(X^*)$ is positive definite
In case of a single variable function, the conditions becomes

$$f'(x^*) = 0$$
$$f''(x^*) > 0$$

C.7 Newton Method

Newton method is a derivative based method. It is based on exploiting the quadratic approximation of the function to be minimized. The quadratic approximation $q(x)$ of the function $f(x)$ at the point x_k is

$$q(x) = f(x) + f'(x_k)(x - x_k) + \frac{1}{2}f''(x_k)(x - x_k)^2$$

The next point in Newton method is taken to be the point where the derivative of the quadratic approximation vanishes.

$$x_{k+1} = x_k - \frac{f'(x_k)}{f''(x_k)}$$

The method terminates when $|x_{k+1} - x_k| < \in$, or when $|f'(k)| < \in$, where $\in$ is a pre specified termination scalar.

C.8 Hooks and Jeeves Method

The method of Hooke and Jeeves performs in each step two types of search. The first type is an exploratory search along the coordinate directions and second type is a pattern search along the direction $X_k - X_{k-1}$, where X_k is a point obtained after performing iteration k of the algorithm. The original method of Hooke and Jeeves performs discrete steps along the search directions. Here the continuous version of the method using line search along the coordinate direction $d_1, d_2, \ldots, d_n$ is presented

Step 1 Choose a scalar $\in > 0$ to be used as stopping criterion for the algorithm. Choose a starting point X_1. Let $y_1 = X_1$, let $k = j = 1$ and go to step 2

Step 2 Find the optimal λ_j for the problem minimize $f(y_j + \lambda d_j)$ and let $y_{j+1} = y_i + \lambda_j d_j$. If $j < n$, replace j by $j + 1$ repeat step 2. Otherwise if $j = n$ let $X_{k+1} = y_{n+1}$. Go to step 3

Step 3 If $\|X_{k+1} - X_k\| < \in$, stop, otherwise go to 4

Step 4 Let $d = X_{k+1} - X_k$ and let λ^* be the optimal solution that minimizes $f(X_{k+1} + \lambda d)$. Let $y_1 = X_{k+1} + \lambda d$. Let $j = 1$, replace k by $k + 1$ and go to step 1.

References

1. Blank L (1980) Statistical procedures for engineering management and science. McGraw Hill Book Company, New York
2. Walpole RE, Myers RH (1993) Probability and statistics for engineers and scientists, 5th ed. Macmillan Publishing Co., New York
3. Montgomery DC, Runger GC (1994) Applied statistics and probability for engineers. Wiley, New York
4. Law AM, Kelton WD (1991) Simulation modeling and analysis, 2nd edn. McGraw Hill International, New York
5. Jardine AKS (1973) Maintenance, replacement, and reliability. Pitman Publishing Corporation, New York
6. Kapur KC, Lamberson LR (1977) Reliability in engineering design. Wiley, New York
7. Bazaraa MS, Sherali HD, Shetty CM (1993) Nonlinear programming theory and algorithms, 2nd edn. Wiley, New York
8. Luneberger DG (1973) Introduction to linear and nonlinear programming. Addison-Wesley Publishing Co., Boston

S.O. Duffuaa and A. Raouf, *Planning and Control of Maintenance Systems*,
DOI 10.1007/978-3-319-19803-3

Index

S.O. Duffuaa and A. Raouf, *Planning and Control of Maintenance Systems*,
DOI 10.1007/978-3-319-19803-3

MIX
Papier aus verantwortungsvollen Quellen
Paper from responsible sources
FSC® C105338

If you have any concerns about our products,
you can contact us on
ProductSafety@springernature.com

In case Publisher is established outside the EU,
the EU authorized representative is:
Springer Nature Customer Service Center GmbH
Europaplatz 3, 69115 Heidelberg, Germany

Printed by Libri Plureos GmbH
in Hamburg, Germany